Chatbots - The AI-Driven Front-Line Services for Customers

Edited by Eduard Babulak

Published in London, United Kingdom

Chatbots - The AI-Driven Front-Line Services for Customers
http://dx.doi.org/10.5772/intechopen.104076
Edited by Eduard Babulak

Contributors
Marina Escobar-Planas, Emilia Gómez, Vicky Charisi, Carlos-D Martinez-Hinarejos, Isabelle Hupont, Matej Rojc, Umut Ariöz, Valentino Šafran, Izidor Mlakar, Aida Huerta Barrientos, David Felipe Mares Silva, Eduard Babulak, John Kontos

First published in London, United Kingdom, 2023 by IntechOpen
IntechOpen is the global imprint of INTECHOPEN LIMITED, registered in England and Wales, registration number: 11086078, 5 Princes Gate Court, London, SW7 2QJ, United Kingdom

British Library Cataloguing-in-Publication Data
A catalogue record for this book is available from the British Library

Additional hard and PDF copies can be obtained from orders@intechopen.com

Chatbots - The AI-Driven Front-Line Services for Customers
Edited by Eduard Babulak
p. cm.
Print ISBN 978-1-83768-931-6
Online ISBN 978-1-83768-932-3
eBook (PDF) ISBN 978-1-83768-933-0

Meet the editor

Professor Eduard Babulak, Ph.D., is an accomplished international scholar, researcher, consultant, educator, professional engineer, and polyglot with more than 30 years of experience. He has been an invited speaker at prestigious academic institutions worldwide, including the University of Cambridge, England; Massachusetts Institute of Technology (MIT) USA; Purdue University, USA; Yokohama National University, Japan; Penn State, USA; and many others. His biography was cited in the Cambridge Blue Book, Cambridge Index of Biographies, and several editions of *Who's Who in the World* and *Who's Who in America*.

Contents

Preface

The emergence of artificial intelligence (AI) is changing the way business processes and customer services are executed in organizations both internally and externally. For example, today, instead of interacting with a human representative, customers are often directed to AI-driven software agents called chatbots, which provide services automatically based on the client's questions. The integration of chatbots in daily business has become a reality and is just the beginning of AI being applied to businesses and many other economic sectors worldwide.

This book presents the latest advances in chatbots and brings to light the latest findings and future research in chatbot architectures. It also explains how chatbots work in different domains, their current technological limitations, and future trends of the technology. The book promotes the creation of interdisciplinary research teams to push the edge of innovation and research of the next generation of AI-driven conversational technologies.

Eduard Babulak
National Science Foundation,
Alexandria, VA, USA

Chapter 1

Introductory Chapter: Journey to AI Driven Chatbots

Eduard Babulak

1. Introduction

Business communication today is driven by the organizational financial performance and customer best quality service provision. The rise of 24/7 ubiquitous access to Internet and Applied Artificial Intelligence (AAI) creates a platform for future fully-automated cyberspace. Role of Computing has become an essential part of research, innovation and development of future cutting-edge technologies that will transform the way we live to a next level, where Machine-to-Machine Communication will drive Human-to-Human Communication.

Today we see the technology that has brought the Computational Machines much closer to us people, and LEXA is becoming an integrated part home companionship and the entertainment for many young families who are tech-savvy and love to play with the high-tech. Naturally, there are many dimensions to technological innovation and application of AI that are fascinating and inspire scholars and industry developers to push the edge of advancement of next generation technologies that will promote a Cyber Automation and Ultra Smart Living for everyone.

The Machine learning (ML) as a subdomain of artificial intelligence enables smart computational device to abstract patterns from data without explicit programming. In addition to AI and ML, we see the rise of Humanoid Robotics that has potential to act like, look like, talk like and reason like human, and eventually becoming an essential part of an organizational business infrastructure worldwide [1].

Some experts compare level of intelligence of new Humanoid Robot called Sophia to that of a one month old new born baby that is at the very beginning of discovering the world around and is beginning to observe and to learn to communicate with his or her parents. Yet, very young and tiny, babies have a Natural Intelligence and Human Reasoning that may be compared to a computational capacity of most performant Supercomputer in the world today.

Many experts suggest that there are fundamental differences between the AI and Natural Intelligence (NI) and that future evolution of AI driven Software Agents and Humanoid Robotics may have strong impact on the future Industry 5.0 and 6.0, Business to Business (B2B) and Mobile Commerce (mC). Given recent COVID relates crisis, applications of AI in business, industry, government, academia and other sectors have become ubiquitous showing more and more examples of Smart Software Agents application.

The next section, presents examples of the AI driven applications including chatbots and ChatGPT, and discus what, how and where they are.

2. What are the Chatbots?

Chatbots are one of the practical examples of AAI driven cyber automation. The software capable to facilitate human-machine conversation is known today as a chatbot. A chatbot software simulates and drives a real-time interaction between a customer a human partner, while bridging a client system and information acquisition domains.

Chatbot is a conversational agent that utilizes AI to interpret the text of the chat using Natural Language Processing (NLP). Instead of direct communication with human service personnel, the customer/client can make conversation via text or voice with the chatbot software [2].

The chatbots utilize the AI embedded conversational systems capable to recognize customers and information systems database driven keywords or word patterns to provide the human-like answer in real-time. Chatbots have become an essential part of organizational business infrastructures while improving a customer service with faster and cost effective best quality support.

2.1 Chatbots are not perfect yet

The rise of AAI and chatbots have created collection of new tools that are often applied by students, faculty and others to write scholastic essays, articles, reports and books. Given the AI vs NI capabilities, there are many remaining technological challenges, including:

- Human to Human Perceptions: Due to lack of human like NI, the AI and chatbots are capable of generating the text based response on client's data with certain limitations and to compose and write machine generated response only. Human to human like perceptions are important part of effective communication and it remain to be challenge for the AI and chatbots developers.

- Emotional intelligence (EI): Having clarity on what and how human emotions impact the composition and effective writing is essential part to communicate and convey the ideas effectively. Today, the AI and chatbots can generate text that is grammatically correct and factually accurate, but are unable to compute a human EI, which is natural for humans.

- Contextual understanding: The current AI and chatbots are processing information system database data without capabilities to understand a broader context in which ideas are communicated in writing, which ultimately generates errors and inaccuracies in the machine generated text.

- Intuition and Creativity: AI and chatbots today are capable to assist clients with the text generated by organization information system database data only, which does not reflect the level of human intuition and creativity. There are many technological challenges for AI & chatbots to generate original human like ideas and to communicate seamlessly the way humans do.

- In light of current advances in the field of AAI and chatbots, there are great improvements in machine assisted composition, rising to level close to Human like capabilities to write and to communicate the ideas effectively. The role of humans will continue to be essential in the process of innovation and development of future more sophisticated AI driven chatbots.

2.2 New generation of Chatbots

Evolution of AI driven chatbots brough a new Chat Generative Pre-Trained Transformer (ChatGPT) technology utilizing a dialogue-based AI chatbot capable of understanding natural human language and generating quite detailed human-like written text. These ChatGPTs are part of text-generating AIs technologies and are becoming integrated in business, industry, government and other sectors worldwide.

The Chat GTP systems are trained by AI & ML, and are capable to answer questions via a conversational interface. These new OpenAI systems are trained on a very large sample of text adopted directly from the Internet information System, enabling a dialogue format for ChatGPT to answer or follow-up questions, to admit its mistakes, to challenge incorrect premises, as well as to reject inappropriate requests [3]. The results produced by ChatGPT are based on the data available in the organizational information systems database only. The ChatGPT today, are not able to collect data from primary sources independently and to advance the AI driven technological evolution in social sciences.

There are important challenges to make ensure data integrity and data accuracy in conjunction with the structural and legal frameworks regarding the copyright and the author's rights. There are critical challenges to make sure that the future impact of AI driven chatbots on mental health is minimum.

Current studies show, that the anxiety and paranoia levels have increased when interacting with chatbots acting like humans instead of humans communications with humans. Some social applications may mislead people with chatbots acting like humans to help people to socialize utilizing chatbots acting like humans. This may contribute to paranoia, and some users may not be able to distinguish whether they are talking to humans or chatbots. The studies show that extensive use of ChatGPT may contribute to paranoia caused by frequent interactions with chatbots instead of interaction with humans [4].

3. The Chatbots applications in B2B

Today, the chatbots are well accepted in the B2B providing a business value and promote framework for training chatbots in order to to better serve B2B customers.

Chatbots have potential to provide support for fast and effective B2B shopping experience by providing personalized support and guidance, helping customers to find the right products or services quickly and efficiently. The B2B chatbots are trained to understand the unique client's needs and preferences, by utilizing a data analytics tools to seek insights into customer experience and preferences. By leveraging this data, chatbots can be trained to provide personalized recommendations, answer specific questions, and guide customers through the purchasing process [5].

One of the most essential criteria for a successful business is a customer satisfaction and perception of having the best quality of service with the best quality and guarantees of company services or products. The business to client and client to business communication is fundamental part of the company's success and future sustainable and successful growth. The next section, bring to light future challenges that chatbots will be facing in the years to come.

4. Chatbots future challenges

With continuing increase number of Chatbots in the government, business, academia, industry, and other domains, there is a new trend to market their services and products via chats and bots 24/7 worldwide. Given the large popularity of utilizing the chatbots applications, there are technological challenges concerning the impact that different usage contexts have on the chatbots' application in mobile commerce (mCommerce). Given number of differences in the nature of mobile business not all shopping contexts may be best fit for chatbots.

To address these challenges the following chapters present various examples of the client's perceptions and level of adoption of chatbots in mobile commerce. The current studies show that Chatbots are more suitable in the context of one-attribute, information-light, and group-buying tasks, whereas traditional Apps are suitable for multi-attribute, information-intensive, and single-buying scenarios [6]. There are yet new opportunities to design novel chatbots that will provide best user experiences and ultimately enhance the user perceptions and adoption intentions in all commercial sectors. The path to future is often found while looking in the past and asking a simple question, how did we get here [7–9].

Author details

Eduard Babulak
National Science Foundation, Alexandria, VA, USA

*Address all correspondence to: babulak@yahoo.com

References

[1] Babulak E. "Blog Article:", The Third Millennium AI-Driven Humanoid Robot. Switzerland: SwissCognitive, World-Leading AI Network; 2023

[2] Zerouani H, Haqiq A, Bounabat B. Chatbots technology and its challenges: An overview. In: International Conference on Advanced Intelligent Systems for Sustainable Development. Volume 1 - Advanced Intelligent Systems on Artificial Intelligence, Software, and Data Science. 2023. ISBN: 3031263839, 9783031263835 DOI: 10.1007/978-3-031-26384-2_6

[3] Web source: https://www.theguardian.com/technology/2022/dec/05/what-is-ai-chatbot-phenomenon-chatgpt-and-could-it-replace-humans [Accessed June 28, 2023]

[4] Uludağ K. Talking to Chatbot or Human Paranoia. 2023. 10.13140/RG.2.2.27265.63843

[5] Ghildiyal V, Adnan R. Enhancing B2B Customer Experience with Personalized Chatbots. 2023. 10.13140/RG.2.2.35821.36323

[6] Hsu P-F, Nguyen K, Wang C-Y, Huang P-J. Chatbot commerce – How contextual factors affect Chatbot effectiveness. Electronic Markets. 2023;**33**(1):1-22. DOI: 10.1007/s12525-023-00629-4

[7] Babulak E. Future computing and Cyber Home 2015. In: International Review on Computers and Software (IRECOS). 2006. Available from: http://www.praiseworthyprize.com/IRECOS-latest/IRECOS_vol_1_n_1.html#The_Future_Computing_and_CyberHome_2015

[8] Babulak E. Information cyberspace and future cyber homes. International Journal of Online Engineering (iJOE). 2006;**2**(2). DOI: 10.3991/ijoe.v2i2.338. Available from: http://online-journals.org/index.php/i-joe/article/view/338

[9] Babulak E. Interdisciplinarity and ubiquitous internet technologies in support of automation. International Journal of Online Engineering iJOE. 2006;**2**(2):1-6. Available from: http://www.online-journals.org/index.php/i-joe/article/view/304

Chapter 2

A State-of-the Art Survey on Chatbots Technology Developments and Applications in Primary Healthcare Domain

David Felipe Mares Silva and Aida Huerta Barrientos

Abstract

Chatbots, defined as artificial intelligence program able to simulate processes of human conversation *via* auditory or textual methods, are deployed by firms to automate customer service. In recent years, chatbots have received tremendous attention from scholars in numerous fields including e-health, e-learning, and e-commerce over many sectors. However, the technology developments and applications specifically in the primary healthcare domain are still insufficiently explored. The principal purpose of the study is to provide a broad review of the current technology developments and applications in primary healthcare domain and future directions in the research. First, we describe features of chatbots considering the healthcare domain. Next, we provide a classification of technology developments and applications in primary healthcare with a focus on recent advances. Then, we present a density map of applications in the primary healthcare domain. Furthermore, we introduce future directions in the core research technology. We expect this study to serve as a comprehensive resource for researchers in healthcare domain.

Keywords: artificial intelligence, chatbot, technology development, primary healthcare, ChatGPT

1. Introduction

The changes in the epidemiological gradient that has been observed in the last decades together with the digital transformation have affected different areas of health that, added to the trends toward health prevention, well-being promotion, and personalized medicine, are generating a change in health systems to move away from hospitals and bring services closer to people and their homes [1, 2]. This new perspective that seeks to integrate health systems around people, instead of diseases, has influenced different sectors such as education, infrastructure, communications, finance, economy, and work. Thus, this has been the object of interest for health applications of the technological advances achieved by new business and industrial application models, around different spheres of hardware, software, connectivity, and services and new technologies such as artificial intelligence (AI) and Health 4.0 [3].

Within the technological developments associated with artificial intelligence and mobile communication systems applied to healthcare, chatbots represent a trend that is increasing in popularity as an efficient mechanism that promotes interactions between application users for different sectors, since it provides personalized information and allows interactions in time and a capacity to reach millions of people at the same time [4, 5]. From the patient's perspective, chatbot technologies as representation of natural language processing, along with deep learning and virtual reality, also referred as cognitive services, have been identified as healthcare drivers by their possibility for the creation of great impact applications on medical and preventive health services [2, 6].

An important segment of the technological advances in health in recent years has focused on the use of artificial intelligence, telemedicine, and automated monitoring of physiological signals for the benefit of patient-centered medicine under a vision of personalization and digital well-being, characterized by being preventive, personalized, predictive, and participatory [7]. This personalization of medicine has been promoted through health policies that seek to expand access to primary healthcare (PHC) as a performance indicator for healthcare systems, since it aligns public policy and service provision at the individual level through healthcare services and primary care and at the population level through public health [2, 8].

Chatbots, defined as artificial intelligence programmable to simulate processes of human conversation *via* auditory or textual methods, are referred to in the literature in various ways, such as conversational agents, embodied conversational agents (avatars), social robots, dialog systems, voice user interfaces, and voice assistants, all of which mimic human conversation using text and/or spoken language [3, 5, 9]. In the healthcare domain, chatbot applications have shown good results for performing repetitive tasks of low complexity, delivering personalized content that allows patients to gain greater insight into their health conditions, and have shown the ability to improve patient engagement in certain contexts [10]. So, different applications are observed in healthcare, ranging from mental health, assisted living, customer service, support in states of depression, substance abuse disorders, filling in clinical history, nutrition recommendations, diet, exercise, evaluation of respiratory symptoms, support in the administration, and supply of medications [2, 3, 11]. Many benefits in the field of healthcare are derived from the chatbots' capabilities to be continuously available with up-to-date information, hear and respond in natural language, being able to present information in local languages and dialects, reach millions of people at the same time, supporting environments where health professionals are scarce to off-load repetitive tasks that absorb the time of health professionals, as well as anonymity protection for sensitive health condition [12]. In fact, this technology is still considered in a state of initial development due to the implications derived from the technology such as medical dilemmas in its use such as the lack of empathy perceived by the users, the complexity in the interaction of patients' beliefs about diseases that impact the acceptability and the responsibility of chatbots, content quality, accuracy, sources used, patient safety, and diagnostic capacity, among others. From the quality of chatbots perspective, there are differences not only in the technology, interface, contents, and applications, but also in the methodology for measuring both quality and efficiency [13].

According to Refs. [14, 15], the chatbot applications have proved to be useful for public health functions to deal with the COVID pandemic, by encouraging the adoption of strategies of promotion, prevention, mass dissemination of information, reduction of misinformation as it was used by governments and by the World Health

Organization (WHO) to prepare collective response actions. Despite the fact that there were more than 300,000 health applications available globally in 2017 that were available for download, the measurement tools and evaluation studies of the aspects surrounding the provision of health services through chatbots are still very small [13].

Besides, there are aspects that require further study and supervision for the correct use of chatbots in health-related environments, since chatbots use demographic data of patients that they collect through interactions and that may have legal and ethical complications. This could be the case when chatbots, which do not have the appropriate corpus and are not ethically framed and supervised under good practices by a health professional, may end up presenting risky responses for sensitive conditions such as the use of substances, the combination of medications, or mental health advice [2, 12, 16]. This is added to the fact that the chatbots' programming can behave like a black box with answers that are difficult to follow in its construction, presenting logical but not necessarily correct results [4, 7].

As suggested by Refs. [4, 7, 17], the developers of the substantive processes of the chatbots or the intelligence behind it may have trained said chatbots with incomplete information, with empty data, missing data, opaque imputation rules, or based on target populations that are not necessarily generalizable. Likewise, users of these systems can enter inaccurate information due to the inherent teleology of each person who uses it. For such reason, special consideration should be given to developing evaluation methods of different aspects related to chatbot training for health, not only to measure its usability or acceptability by the user. The ethical and clinical dimensions should be measured as well, which also implies expertise and clinical experience, since these will adapt the best response to the patient's clinical problems, according to their characteristics and needs and, of course, always ensuring their clinical safety [4, 18].

A small group of studies has been identified to measure dimensions associated with health. These present a wide variety of methodologies, sample widths, randomness, and population stratification samples to carry out the measurements with the purpose of evaluating these dimensions [10, 12]. Various use scenarios have been visualized where great benefits can be obtained for health systems in terms of efficiency, among which the capacity to process large volumes of patients stands out, where the risk of late diagnoses must be balanced with the use of resources [9, 19].

This chapter is organized as follows. We begin by reviewing the characteristics and features of chatbots, introducing generalities about primary healthcare in Section 2. The technology developments and application domains in healthcare are presented in Section 3, focusing on the primary healthcare public policy-oriented applications. A revision of benefits, challenges, and trade-offs for healthcare delivery and value-based care models is presented in Section 4. Based on these discussions, some future directions are outlined in Section 5, and we give our general conclusion in Section 6.

2. Features of chatbots considering healthcare domain

The WHO [20] defines PHC as *a whole-of-society approach to health that aims at ensuring the highest possible level of health and well-being and their equitable distribution by focusing on people's needs and as early as possible along the continuum from health promotion and disease prevention to treatment, rehabilitation, and palliative care, and as close as feasible to people's everyday environment*. In this direction, as recommended in Ref. [8], PHC is a combination of public health, medical care, and social assistance, which is essentially anticipatory in nature, and includes actions in different aspects such as health education, food and nutrition, environmental sanitation, maternal-child health, immunizations, prevention and control of endemic diseases, treatment and control of the most frequent diseases, and provision of essential medicines. Under the PHC policy, this analysis of the impact of environmental, physical, demographic, epidemiological, congenital, economic, and social factors is essential for the provision of person-centered health services, taking care to maintain the focus on the prevention and promotion of health before it becomes disease.

In the context of personalization of healthcare, service value-based care models are expected to use persons and patients context, to improve multiple health conditions of patients and populations [7]. Within cognitive computing applications, developments with multiple uses are identified. As Improta et al. [21] explain, a patient can be used both as a decision support system for medical specialists in the phases diagnosis and treatment and as a monitoring system of the clinical environment in health establishments. The use of these cognitive systems also encompasses digital therapies focused on dream therapy [22], the use of a gaming approach for depressive events [23], and treatment of depression and anxiety [24] of both the young and the elderly [25]. AI-based chatbot systems, due to their characteristics of acting as automated conversation agents, play a central role in various health actions, since they can promote health, by providing education and potentially causing behavioral changes. This is observed in the treatment of adolescents with a chronic medical condition using a text messaging platform (chatbot) with written interactions to increase engagement and deliver educational content [26].

On the one hand, the first contact functions, whether they are collective actions or toward the person, usually begin with the exchange of information between people and health personnel, whether they are doctors, nurses, or health promoters, which is essential for developing action plans for healthcare. Such data collection and its corresponding registration in the institutional systems take time, which, if automated and systematized, could increase the effective consultation time for patient management [2]. In this direction, chatbots interact with patients for specific, short, repetitive, and massive tasks. Rule-based chatbots represent great potential for prevention and health promotion tasks [27].

On the other hand, the use of chatbots that interact with patients through natural language processing, can, in addition to obtaining information from patients, perform iterative data collection based on previous responses to build clinical histories and contexts of health conditions [14]. So, health service providers, whether public institutions or private providers, can strengthen their technological instruments with chatbots that perform these basic tasks of collecting information or disseminating healthy practices and training in self-management of people's health. Thus, the information built can be used both for the personalization of responses for patients and for guidance and promotion on services of greater complexity or specialization required by people.

A fine-trained chatbot that includes capabilities for consultation, knowledge gathering, basic reasoning, and giving feedback can accomplish this guidance task, simulating a health professional. Development frameworks that integrate question-and-answer reasoning mechanisms based on a domain-specific knowledge base can achieve this [28]. Chatbot applications in health, in addition to primary care, have covered marketing and research topics. Such is the case of pill reminders, interviewing smoking habits, dietary behavior, and physical activity and even for extraction system to extract mentions of adverse drug reactions from the highly informal text in social media. Also used are voice agents for chronic illness monitoring, medical counseling and education, clinical decision support systems (CDSS) for diagnosing infection diseases, and assisting medical personals in diagnosing internal conditions for patients based on larger collection of hospital case records [28]. These agents can offer a wide range of problem-solving functions that can integrate multiple tasks from natural language understanding and knowledge base query to reasoning and giving feedbacks, through an iterative inquiry process. Currently, there is a wide range of applications for chatbots aimed at mental health, health education, maternal care and sexual and reproductive health, nutrition and physical activation, sleep disorders, support in emergency situations for chronic diseases, management of respiratory diseases and accidents, increase in self-care, and transitions in stages of risk for vulnerable groups [13, 26].

It is important to note that over the past few years, different categorizations of conversational agents have been developed including both text chatbots and voice chatbots. Some also consider the channel on which they are used, whether *via* smartphone, web, or some additional platform where it is used. They can also be classified by the objective they fulfill, whether they are aimed at a function or a specific general-purpose task [5, 29]. However, as suggested in Refs. [5, 30], it is more common to classify the logic approach of the dialogue management system that interacts with user input using a knowledge database to determine the action to be taken in the conversation flow.

Recent literature review studies have observed that the large majority of mental-health oriented chatbots currently in existence do not use machine learning at all, favoring more stable and predictable techniques such as rule-based modeling [2, 31]. However, findings have been presented that indicate that the perception of some users generates a lack of expectation that they will reach a state of development where they will displace the work of health personnel [12]. The most used chatbots in health applications are rule-based; they use a decision tree on a specific condition to define the rules on which the chatbot carries out the flow of the conversation, choosing how it responds to each user input. So, the complexity and resolution of the chatbot depend on the programming logic and the complexity and depth of the rules with which it has been defined. For this reason, these chatbots cannot learn from user conversations or interactions and are limited to the scenarios for which they have been programmed [27]. These chatbots are also part of the dialog systems known as specific task or closed domain that manages to perform tasks specific to a domain such as technical assistance and customer service [29].

Knowledge base-based chatbots use structured data sources that contain knowledge of a specific function (such as frequently asked questions or FAQs) to make that information accessible to users and deliver relevant content. In this way, this type of chatbots uses keywords and functions connected to a knowledge base of multiple databases and data sources [30]. These application-specific domains usually have limited availability of training data. They include linguistic-based approaches, where the user's questions are converted from natural language into a database query, and the identified answer is presented to the user [28].

In dialog systems, the knowledge-based chatbots are known as open domain's retrieval base models since they select a response from a previously constructed repository. This is the complement of generative models, which produces new responses [29]. The open-domain dialogue systems require large amounts of data to train and often allow effective chatbots to be created, even though they cannot perform effectively on specific tasks, and often cannot query databases or add useful information to their chatbots' answers [32]. Nevertheless, information retrieval, knowledge base or NLP, and systematic literature reviews have classified the techniques, algorithms, frameworks, and tools observed as a combination of the one or more of these technologies: Deep Neural Network, Graph Based Lemmatization, LSA, Multi-Document Summarization Naive Bayes, Named Entity Recognition, Parser, POS Tagging, Relation Finding (Similarity Distance), Shallow Syntactical, Stemming, Support Vector Machine, Text Chunking, and Tokenization [30]. For open-domain neural dialog generation, methods are categorized and examined as a variety of main categories such as Reinforcement Learning (RL), Hierarchical Recurrent Encoder-Decoder (HRED), Generative Adversarial Networks (GAN), Variational Auto-Encoder (VAE), Sequence to Sequence (Seq2Seq), and Pre-training Model [29, 30].

For the chatbots that have been built based on machine learning and AI, the first developments used single-layer representations that were appended through the use of word vectors applied for task-specific architectures. Subsequently, recurrent neural networks, RNNs, were used that increased the number of capable representations including context analysis to achieve better results in architectures for specific tasks, until recent advances based on pre-trained recurrent or transformer language models such as ChatGPT, which no longer uses specific architectures [33].

2.1 Healthcare applications uses, evaluation, and acceptance

Paradigm shifts from the medicalized approach to seeking health systems that focus on the patient and not on the disease are driving the need to readjust the structure of health systems to improve access to health services, not only for a more diverse population but also for systems that are aware of individual differences and people's health contexts. This implies that through these technological advances, the health needs of each person in their context and based on their social determinants are the ones that predominate in the access criteria used by intelligent health systems [34]. This implies that each dialogue system development focused on the different health conditions, especially those that are linked to each other and multifactorial conditions, must work together to offer healthcare services supportive complementary technological and clinical personalization that allows offering a robust experience to people who seek healthcare services. However, before their incorporation into health systems and routine clinical practice, it is essential to review the effectiveness of these technologies, in such a way that there is a clear understanding of under which contexts these tools can be used. This includes understanding the frame of reference; technological, ethical, and clinical evidence; and adaptation to specific populations, among others, which must give support and certainty to the developments, as well as understanding the limitations, biases, good practices, evaluations, and contexts of use [13].

The adoption of chatbots integrates different actors and functions within health systems. Health professionals have systematized the data collection, the appointment schedule, and the dissemination and training of patients for self-care of health and the increase of health literacy. These have proliferated in mental-health and primary-care applications for low complexity actions. Health units are exploring the capacity of chatbots for functions of health education and counseling support, assessment of symptoms, and assistance with tasks such as patient intake process, scheduling, and collecting personal and family histories [2, 28].

Health systems have combined the chatbots with decision support systems to prioritize targets during the pandemic, assess drug side effects from electronic medical records, disseminate information on available health resources, and Management

DOI: http://dx.doi.org/10.5772/intechopen.110847

of installed capacity in high demand situations [2, 28]. The evaluation of chatbots and the dialogue generation system (DSG) is still in early stages, and the evaluation methods are incipient, even though the conversational AI global market size is expected to grow at high rates, increasing the value by 2025 [35]. Thus, there are research lines on adaptations of automatic metrics to evaluate the responses generated, such as bilingual evaluation understudy (BLEU), Recall-Oriented Understudy for Gusting Evaluation (ROGUE). In addition, human evaluations and combinations between them have been used. Even given the different architecture and logic configurations in the dialogue management systems that govern chatbots, humans have been used to assess consistency, fluency, coherence, and informativeness [29, 32].

Different works have been carried out to categorize chatbots using their health context and core features, as well as their NLP capabilities. The type of user targeted, personalization, data acquisition for implicit or explicit personalization, domain areas of health, theoretical and therapeutic support, security, and privacy also have been studied [2]. In the evaluation of chatbots, some aspects have been described by authors such as if chatbot is programmed to support people, patients, health professionals in tasks. Some studies include satisfaction surveys with a Likert scale as well as measures of the interactions between chatbot acceptability, perceived symptom severity and stigma [3, 12]. There are also other technical characteristics linked to other aspects of the system such as the content, the user interface, the channel of use, and functionalities that are evaluated such as irrelevant answers, frozen chats, and messages in non-readable linguistic structures that make them nonfunctional [2].

2.2 Acceptance

Chatbots may be useful for sensitive health issues in which disclosure of personal information is challenging, since Chatbots were seen as least acceptable as a consultation source for severe health issues, while the acceptability was significantly higher for stigmatized health issues [2]. There are studies that explore the use of chatbots, smartphones, text messages, and social networks to provide tools and resources to help in psychological transitions, more frequently oriented to specific age groups (adolescents), where the motivations for using health chatbots are explored, in order to predict their acceptance [36]. Some studies report difficulties for chatbot users to understand how they work, suggesting that there may be difficult concepts to understand and that their acceptability may depend on different aspects such as expectations, favorable conditions for their use, social influence, habits, associated costs, and even access to the health system [12].

Some works point out the concerned the ethical implications as the main obstacle to the adoption of these technologies in the treatment of addictions. Some of these are using a nonhuman agent in a supportive role, giving answers contrary to the intention of the users, giving sensitive information to enhance the effect of medications through explicit indications, or even the potential for causing harm to specific populations [3, 18]. Although many countries are developing and using chatbots as app interfaces focused on treating health conditions, during the COVID pandemic, different coordinated efforts were made between industry, governments, and nongovernmental actors to integrate communication strategies to reach millions of users. This was achieved by different approaches and uses to collect and disseminate information on patients, the virus and the situation of health systems, diagnostic support, guidance on conversions of health systems, and vaccination strategies and carry out telehealth actions [14, 37–39].

It should be remarked that technological advances toward precision-driven healthcare, which promotes the application of data science, in particular technologies, such as interactive cognitive systems, artificial intelligence, and machine learning, are aimed to enhance healthcare provision, to solve the patients' personalized demands more accurately and, at the same time, more easily to the service providers. In this direction, many chatbot technologies are still to integrate the health condition-monitoring continuum since they are still task specific, by health condition, environment, or agent [7].

3. Technology developments and application domains in primary healthcare

This study adopted a systematic literature review approach to achieve the set out objectives per the Preferred Reporting Items for Systematic Reviews and meta-Analyses (PRISMA) approach and guidelines provided by Kitchenham [40]. PRISMA is employed because it is *evidence-based*; the steps involved are auditable and have been well established and used in the literature and similar studies [41].

3.1 Search strategy

The search was conducted in Scopus and Science Direct. These databases were selected because they host high-impact publications relating to the chatbots for PHC domain and have been used in reviews covering similar themes. Also, the search was complemented with a citation-tracking approach, which involves checking the reference list of relevant publications to track other relevant publications.

3.2 Selection criteria

General search criteria: The developed search query was used in Scopus without year restriction with a focus on the subject areas of *Chatbots* and *primary healthcare* and only documents available in the English language.

Refining criteria: Studies were included based on these predetermined criteria:

1. Studies that involved the application of conversational AI in the healthcare domain,

2. Studies that involved the development of conversational AI in the healthcare domain,

3. Studies that involved the integration or evaluation of conversational AI in the healthcare domain.

Similarly, articles were excluded based on these predetermined criteria:

1. Studies that did not employ conversational AI for application development,

2. Review studies that mentioned conversational AI but did not explore its applications.

3.3 Data extraction

The result was further reviewed, and the following information was extracted from each publication:

1. Year.

2. Authors.

3. Title.

4. Publication type.

5. Aim.

3.4 Results

Health systems based on primary care in the digital era are turning toward collaboration and the generation of complementary service partnerships that make it possible to obtain many of the benefits of economies of scale through the segmental participation of each actor and health service provider [42]. These value chains that are being built around the current installed capacity and the capacities of the (human) medical teams seek multichannel strategies to offer complementary services. Among these strategies are those that allow systematized service channels such as chatbots from which it has been observed that customer service advantages are obtained.

Public Health application chatbots across many countries and languages other than English were used to reach millions of citizens and all relevant institutions, in order to deliver education of citizens, surveillance and detection of contacts, risk assessment, and dissemination of information, allowing the organization of health systems while promoting cooperation, community, involvement, and accountability of citizens through collective actions [15, 37–39]. Healthcare Evolution, which includes Health 4.0 through telemedicine and artificial intelligence, is emerging toward customization and models based on value, which are connected by integrity into the business models that accompany people throughout life, by customization of healthcare, based on the optimization of determinants by reducing the follow-up effects of different health conditions [7].

The dimensions such as education, economy, income, finance, food security, communications and transport infrastructure, assurance, access to health services allow to investigate the social determinants of health that make up the context of life of people. This health information from the social context improves the personalization toward the person, by combining these data with the medical and family records, to add the self-care and personalized management, the PHC system's strategies to chronic disease monitoring, and provision of medical and assistance services [9]. As part of cognitive services, expert systems for healthcare were explored in the Scopus database. We obtained 1138 results that are observed in the cluster in the **Figure 1**. Green color presents various topics related to decision making, decision support, algorithms, data mining, machine learning, deep learning, and information classification.

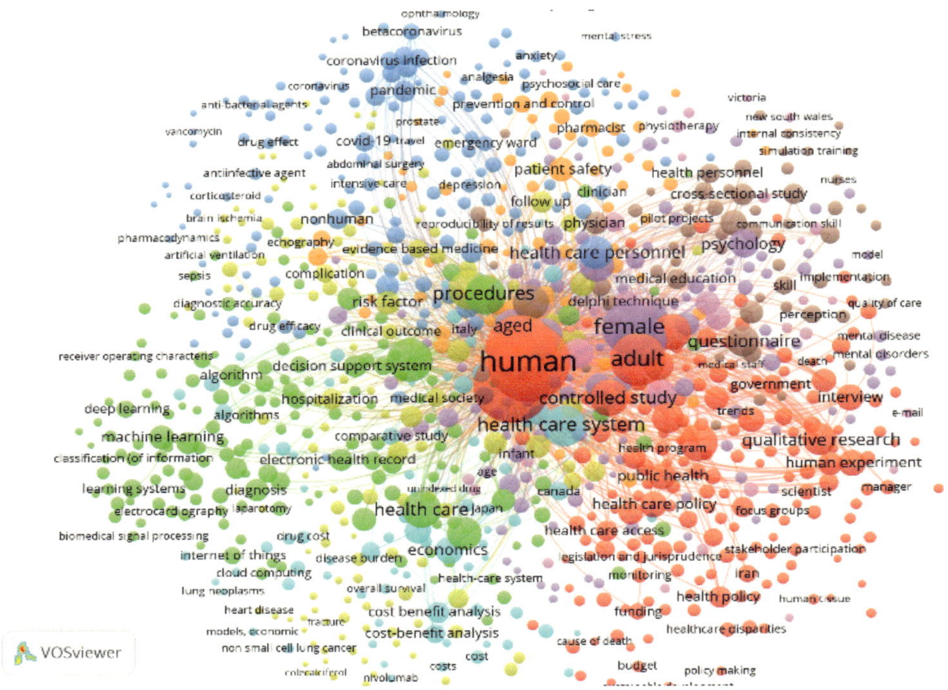

Figure 1.
Visualization by VOSviewer software of word co-occurrence network built using words in titles and abstracts of documents on healthcare expert system field.

Likewise, in the cluster in the color blue, a set of health conditions is presented where the appearance of research related to the SARS-COV 2 virus (COVID-19) is observed and where topics on mental health, anxiety, stress, and depression also appear. In red, public health issues that address different aspects of public healthcare policies are observed, going through issues such as government, financing, budget, participation, inequities in health and access to health, and causes of death. In turquoise, economic topics appear under labels such as economic models, cost-benefit analysis, costs, disease burden, and drug costs. In olive green, the risk themes appear, where risk factors, evidence-based medicine, patient follow-up protocols, patient safety, complications, and hospitalization are displayed. In brown, the nodes with contents linked to existing soft skills in hospital management are presented, such as perception, quality of healthcare, interpersonal and communication skills, and internal consistency. Finally, in pink, there are some small nodes representing studies on issues related to mental health, psychology, and their conditions such as mental illnesses and disorders.

Aforesaid, once again, a close link is observed between the topics where there is no dominant approach that segments the cluster based on a particular research topic. In the same way, for this image, it is interesting to observe that although machine learning and deep learning appear, there are no nodes referring to artificial intelligence. **Table 1** shows the results for PHC chatbot search in Scopus and Science Direct databases. The research topic is still a new research direction, contrasting to the big healthcare apps developed in recent years and available on apps marketplaces.

Weobot, as a self-care expert, trained and tested approaches as cognitive behavioral therapy (CBT), mindfulness, and dialectical behavior therapy (DBT), have reached a significant awareness for the accompaniment of the substance-use

Year	Author	Proposal
2020	[21]	They present a patented device for automatic processing of clinical data of chronic poly-pathological patients.
2020	[22]	This is a proof-of-concept study, which aims to evaluate the feasibility, acceptability, and preliminary efficacy of a digital cognitive behavioral therapy for insomnia (dCBT-I) for individuals with CM and insomnia (CM-I) in the United States.
2020	[43]	They introduce a chatbot architecture for chronic patient support grounded on three pillars: scalability by means of microservices, standard data sharing models through HL7 FHIR, and standard conversation modeling using AIML.
2019	[44]	They proposed a method as a mobile health service in the form of a chatbot for the provision of fast treatment in response to accidents that may occur in everyday life, and also in response to changes of the conditions of patients with chronic diseases.
2019	[36]	This research aimed to explore participants' willingness to engage with AI-led health chatbots.
2019	[26]	The analysis of the use of smartphones, text messaging, and social media prevalent among teenagers, to engage in their preferred channel to provide tools and resources to help them successfully transition to adult-focused care, is presented.
2019	[45]	kBot, a knowledge-enabled personalized chatbot system designed for health applications and adapted to help pediatric asthmatic patients (age 8 to 15) to better control their asthma. Its core functionalities include continuous monitoring of the patient's medication adherence and tracking of relevant health signals and environment data. kBot takes the form of an Android application with a frontend chat interface capable of conversing in both text and voice, and a backend cloud-based server application that handles data collection, processing, and dialogue management. It achieves contextualization by piecing together domain knowledge from online sources and inputs from our clinical partners.
2019	[46]	The study presents case studies in the healthcare industry that focus on the use of Chatbots to improve patient monitoring and medical services.

Table 1.
Results for PHC chatbot search in Scopus and science direct databases.

disorders; nevertheless, there are a wide range of tests to analyze performance to generate advances in drug/alcohol use [3]. IBM's Watson Cloud Services has developed an app that allows chatbots for different applications, including assistants for learning the treatment process in radiotherapy for cancer patients, genomics, measurement of intellectual disability in children, support for depression episodes in older adults, and medical imaging [47].

Likewise, the use of information and communication technologies that enable quick, simple, clear, and unambiguous access to health information also contributes to the quality of health services. But the focus of public health is very different from the individualized practice of clinical medicine, and as such, public health values and ethics have several justifiable challenges that differ from medical or bioethical ones. Public health is aimed at the population, not individuals, and because of its nature, it is interested in public good [15]. Specific collective health and public health applications for large populations could be observed in many governments, civil society, and international organizations such as WHO. To provide verified information, updated news and reports were provided on pandemic as well as details of symptoms and measures to discount public health systems [15, 39].

3.5 ChatGPT

As discussed earlier, NLP facilitates the interaction of human language with computer systems, so with the release of Generative Pre-trained Transformer 3 (GPT-3), the most recent release version of a language model that uses deep learning to generate text similar to human natural language, it has been trained using large datasets [48]. The NLP systems have evolved from single-layer representations of neural networks using word vectors fed to task-specific architectures; later, multilayered recurrent neural networks (RNNs) were used to add context to achieve better representations, and now pre-trained recurrent or transformer language models have brought great progress in tasks such as reading comprehension, answering questions, and textual detailing, among others. This type of language model has the limitation that even though it can be adapted to different tasks, it requires datasets for specific tasks in order to achieve refinement in those tasks [17]. This means that the almost human responses, the ease of use, and the friendly interaction have a weak side on the training set, which since training with Internet content incorporate deviations from the same information, such as gender, racial, and geopolitical biases [33]. However, the potential for use extends to many of the preventive medicine and PHC applications, especially in the management of electronic clinical records where a correct incorporation of this technology is visualized if said technology is fine-tuned by health professionals that restrain risks in its use.

4. Challenges and trade-offs for healthcare

The capabilities that chatbots will have to favorably influence health work, using the power of computing and big data analysis, which will be exploited by health professionals in ways that are yet to be discovered. However, before the maturation and daily exploitation of this technology arrives, the road will have to go through multiple challenges that are experienced daily in health systems and health establishments, where beyond linguistic accuracy, usefulness, updating of clinical knowledge, the accuracy of medical knowledge, clinical responsibility, the domain of the health condition, the ethics of training and evaluation of the algorithm, and the acceptability of the user, among many others, must be integrated into the problems of organization, budget, culture, generation gap, old infrastructure, telecommunications architecture, financing models, and technologies for healthcare.

Although this technology can structurally change the way of delivering health services, the applications for the delivery of services within hospitals are perceived as much more limited. Hospital procedures, clinical practices, and protocols, due to their complexity and real dynamism, especially in highly specialized fields, place chatbots and NPL technology in the role of support.

However, for the PHC, the range of challenges is wide and almost as large as the universe of application possibilities, which are unique in their forced adaptation and integration to different health systems. In the short and medium term, the challenges for chatbots involve fine-tuning technology, the social dimension of the paradigm shift and evolution of tools, regulations, information management, privacy, data collection mechanisms, and so on. In the case of information management, the information of people and its interoperability with information systems, or medical records are examples of the challenges. The restrictions imposed by the COVID pandemic generated new adaptations of conventional therapies toward digital approaches [14].

Some studies have begun to be carried out for other disorders such as those related to gambling, smoking, sex, Internet, and mobile phone, which manifest compulsive behaviors and will be a subject of deep reflection to review the limits of the use of natural language in healthcare chatbots. However, the evidence indicates that chatbots can take on specific steps within an addiction treatment process and that they should be accompanied by the help of an expert, a peer, or a support group that contextualizes the activities within an addiction program recovery [3]. In this way, it is important to point out that the expected benefits contrast with the less-developed sides of the technology related to design, architecture, the opacity of its programming in the internal layers of the RNNs, privacy, the anonymity of people, honesty in communication, or understanding behavior disorders that lead to irrational actions or even self-deception. These deficiencies have been observed in the responses of the most popular open-domain chatbots and dialogue agents such as Siri, Alexa, Bixby, and ChatGPT, which provide counterintuitive output to user questions, such as mishandling simulated patients with suicidal ideation or providing addresses of marijuana dealers in response to questions about how to treat marijuana addiction [3].

In the same way, even with the increase in the use of chatbots, the evaluation of their effectiveness and feasibility are incipient and it requires more evidence, paying special attention to the methodology used to validate [12]. Similarly, the effectiveness of the soft aspects related to communication competence, and ability to understand users are difficult to evaluate and determinants for their safe adoption [18]. The effectiveness aspects imply the need to pay more attention and effort to include measures that ensure the safety of users, people, and patients, especially in diagnostic work and medical treatment. This complexity is also presented in the privacy and security of the data, the medical control and the authentication of the information obtained by the chatbots. Some regionalized studies showed that these chatbots and apps present from medical disclaimers, HIPAA compliant, child online privacy, and protection act [2]. This scalability characteristic that can bring so many benefits in the prevention promotion that characterizes the PHC, in the collective and population contexts, implies the inclusion of adaptation measures of health messages and prevention campaigns to cultural, educational, social, and economic contexts of each population, including the practices of native peoples [15].

5. Future directions

In recent years, technological advances in the health area have been oriented toward the incorporation of artificial intelligence, telemedicine, and automated monitoring of physiological signals as enablers of patient-centered medicine, the creation of value in health services, and the change toward a culture of prevention through digital well-being. Studies by leading consultants anticipate that chatbots will be used as a first-access channel to help navigate all the options available in the health system, leading people toward virtual solutions or traditional services [42].

The possibilities of chatbots for health will be the person-to-person interfaces of new health technologies, from applications focused on health and well-being and wearable IOT devices to monitoring physiological signals, assisted living, digital mental health therapies, social robotics for nursing care [49], personalized drugs and genomic medicine, and the supply chain around the health and well-being needs of people. As the chatbots in the health domain are increasingly used to enable interaction with humans on an emotional level, for example, to comfort and entertain older people, lonely people, or those with dementia based on cognitive services and intelligence, new ethical challenges also arise to review, such as the need to explore new governance models to guarantee principles such as the *welfare principle*, which postulates artificial intelligence systems (AIS) must, above all, allow the growth of the well-being of all [1, 50]. In this direction, artificial intelligence (AI) is beginning to occupy a central place in the design of new therapies and treatments throughout the different dimensions of healthcare. These capabilities and the impact of cognitive computing will be reflected in the outcome of medical care, well-being, and medicine in the healthcare domain [51]. The personalized healthcare services are transforming the healthcare sector toward the integration of recent technological developments under new value-based care models to improve the efficiency of traditional healthcare systems. Technological advances in health have been oriented in recent years toward the incorporation of artificial intelligence, telemedicine, and automated monitoring of physiological signals, and recent studies have focused on integration around the patient with personalization and digital well-being.

Chatbots will be a strong component in the ecosystem of technologies surrounding the patient in this so-called personalized medicine since they are expected to be used as user interface for all these developments, which involves the use of new AI, Internet of Things (IOT), and genomic technologies to promote participatory, personalized, and participative preventive health where care is designed around people and not a place. This behavioral shift from healthcare to healthy aging will require more efficient and productive public health through the use of new generations of communication technology. In addition, health systems and public health needs will drive a shift in hospitals in the future to become a continuum care facilitator and to mentor people's

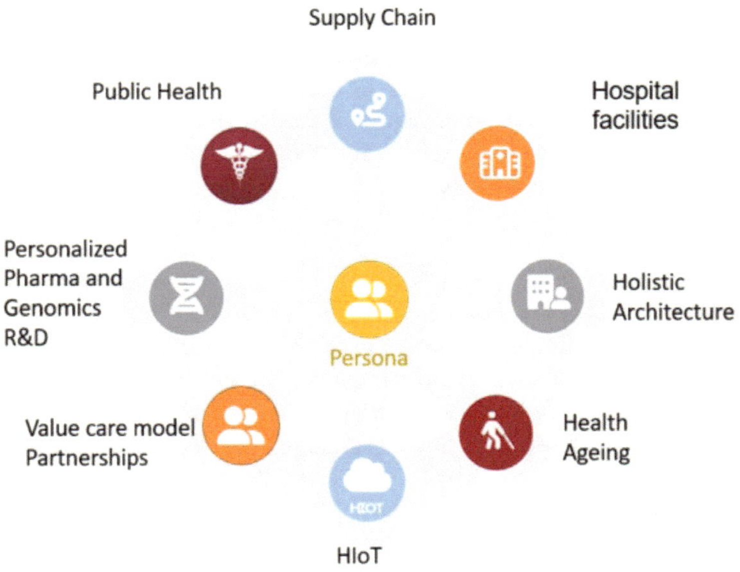

Figure 2.
Health system recent tendencies.

healthy habits. Infrastructure adaptations for holistic health will be pushed by the Internet of Things for health as an enabler of value-based care models (**Figure 2**).

The new pharmaceutical developments toward personalized medicine based on AI are expected to bring personalized medicine closer to patients by articulating integrated healthcare supply chains created around patient experience [42]. Each of these evolutions produced in each health system component will transform healthcare delivery, by forming next-generation integrated healthcare service delivery networks as ecosystems of complementary collaborations around people's well-being. These technologies will allow people to be accompanied through the decisions inherent in each stage of life and throughout the phases of each health condition. Many actions to improve health need to be conducted by other economic sectors and be technology based. Most of the so called *social determinants of health* are beyond the scope of action of health systems; however, health systems need to be prepared to evolve in order to become resilient to social changes, epidemiological shifts, or emergency situations, by proactively detecting early signs of epidemics and be prepared to act early in response to surges in demand for services, which represents a paramount challenge to be faced by disarticulated health service providers. As suggested in Ref. [52], Massive Internet Of Things (MIoT) has been conceived as a viable future scenario to face infectious deseases, especially if combined to other technologies like Blockchain for data privacy and access issues and federated learning. This is conceived as a11 distributed interactive artificial intelligence paradigm, proposed as a solution to single database or big data sets, since it relies on the sharing of machine learning models, instead of the raw data itself, as one of the puzzle pieces of the intelligent healthcare.

Another dimension of this complexity is observed in the training of deep learning models that require large amounts of data, distributed among different institutions and owners, and that, because they contain sensitive medical data, cannot be integrated into a single database in the cloud. For these, new approaches are being explored, such as Federated Learning, a collaborative artificial intelligence technology where information obtained by millions of devices is not shared, but only the trained models [52].

6. General conclusion

Healthbots are potentially transformative in centering care around the user; however, they are in a nascent state of development and require further research on development, automation, and adoption for a population-level health impact [2]. Simple and task-oriented agents will represent a manageable channel for adoption in PHC, for the diagnosis of low-complexity diseases, high stigma, or in contexts where there are few health resources, increasing digital literacy and access. Current trends indicate that chatbots are a mechanism capable of engaging people in healthcare, which allow many interventions to be focused on patients and people; however, there is a lack of a clear regulatory framework for such health interventions, as well as a general opportunity for service providers' active engagement. Similarly, applications that are based on AI should incorporate ways to monitor the measures taken by programmers to ensure the ethical, technical and clinical, and population quality of chatbots along with patient safety in each functionality. Models based on GPT and deep learning will have to mature and refine their responses and the quality of medical information and update their practices for more complex tasks to those searched in the PHC.

Health chatbots have a maturation period ahead, where the development methodologies applied to each use case are standardized. Meanwhile, for primary healthcare, those chatbots that manage to integrate with users, patients, health professionals, and service providers will be the ones that will have an impact on the systematization of repetitive PHC tasks, in which a lot of health systems' time and resources are invested nowadays.

The benefits will reach beyond the limits of health establishments and will go to other applications that, focused on improving the experience of people and patients, promote well-being over disease. In this way, they will impact the continuum of care, prevention, promotion, foresight, and personalization of medicine, together with intelligent collective public health work in real time. What will be seen soon in the healthcare domain for chatbots will be applications that include interaction on social networks, triage of patient symptoms, support and counseling before and after the clinical encounter, and as sources of information and support in organizing tasks, process management, appointments, files, medicines, and supply chain, among many other tasks. However, these benefits could be achieved mainly as a complement to a digital approach for healthcare providers, to aid health professionals and improve doctor-patient communication for low-severity conditions. This approach can be led by primary-care services adopters, whose interdisciplinary teams can jump the adoption barriers linked to culture change and technology transfer.

Finally, the challenge will be to assess the best fit for this technology within healthcare systems' settings, and PHC public policies, and how to manage the best approach to incorporate the daily profiles of health personnel who do not use technology, especially in countries with a lower level of development, which run the risk of not being able to integrate the value base that makes this technology possible, limiting themselves to becoming end users or even staying completely outside the new paradigm.

Conflict of interest

The authors declare no conflict of interest.

Author details

David Felipe Mares Silva and Aida Huerta Barrientos*
Faculty of Engineering, Department of Operations Research, National Autonomous Mexico University, Mexico City, Mexico

*Address all correspondence to: aida.huerta@comunidad.unam.mx

References

[1] Kickbusch I. Health promotion 4.0. Health Promotion International. 2019;**34**(2):179-181

[2] Parmar P, Ryu J, Pandya S, Sedoc J, Agarwal S. Health-focused conversational agents in person-centered care: A review of apps. npj Digit Med. 2022;**5**(1):21

[3] Ogilvie L, Prescott J, Carson J. The use of Chatbots as supportive agents for people seeking help with substance use disorder: A systematic review. European Addiction Research. 2022;**28**(6):405-418

[4] Wang J, Hwang G-H, Chang C-Y. Directions of the 100 most cited chatbot-related human behavior research: A review of academic publications. Computers and Education: Artificial Intelligence. 2021;**2**:100023

[5] Saka AB, Oyedele LO, Akanbi LA, Ganiyu SA, Chan DW, Bello SA. Conversational artificial intelligence in the AEC industry: A review of present status, challenges and opportunities. Advanced Engineering Informatics. 2023;**55**:101869

[6] Muschenich M, Wamprecht L. Health 4.0 - how are we doing tomorrow? Bundesgesundheitsblatt, Gesundheitsforschung, Gesundheitsschutz. 2018;**61**(3):334-339

[7] Taimoor N, Rehman S. Reliable and resilient AI and IoT-based personalised healthcare services: A survey. IEEE Access. 2021;**10**:535-563

[8] Dassah E, Aldersey H, McColl MA, Davison C. Factors affecting access to primary health care services for persons with disabilities in rural areas: A "best-fit" framework synthesis. Global health research and policy. 2018;**3**(1):1-13

[9] Hong G, Smith M, Lin S. The AI will see you now: Feasibility and acceptability of a conversational AI medical interviewing system. JMIR Form Res. 2022;**6**(6):e37028

[10] Seitz L, Bekmeier-Feuerhahn S, Gohil K. Can we trust a chatbot like a physician? A qualitative study on understanding the emergence of trust toward diagnostic chatbots. International Journal of Human-Computer Studies. 2022;**165**:102848

[11] Jang S, Kim J-J, Kim S-J, Hong J, Kim S, Kim E. Mobile app-based chatbot to deliver cognitive behavioral therapy and psychoeducation for adults with attention deficit: A development and feasibility/usability study. International Journal of Medical Informatics. 2021;**150**:104440

[12] Miles O, West R, Nadarzynski T. Health chatbots acceptability moderated by perceived stigma and severity: A cross-sectional survey. Digital Health. 2021;**7**:1-7

[13] Aitken M, Clancy B, Nass D. The Growing Value of Digital Health: Evidence and Impact on Human Health and the Healthcare System. Parsippany, NJ, USA: IQVIA Institute for Human Data Science; 2017. p. 1

[14] Bharti U, Bajaj D, Batra H, Lalit S, Lalit S, Gangwani A. Medbot: Conversational artificial intelligence powered chatbot for delivering tele-health after covid-19. In: 5th International Conference on Communication and Electronics Systems, ICCES 2020. Coimbatore, India: Institute of Electrical and Electronics Engineers Inc; 2020

[15] Petričević S, Mustić D. Communicating a global pandemic

with whatsapp and healthbot in Croatia. In: 11th International Symposium on Graphic Engineering and Design, GRID 2022. Novi Sad, Serbia: University of Novi Sad - Faculty of Technical Sciences, Department of Graphic Engineering and Design; 2022

[16] Belen Saglam R, Nurse JRC, Hodges D. Privacy concerns in Chatbot interactions: When to trust and when to worry. In: Stephanidis C, Antona M, Ntoa S, editors. 23rd International Conference on Human-Computer Interaction, HCII 2021. Springer Science and Business Media Deutschland GmbH; 2021. pp. 391-399

[17] Nath S, Marie A, Ellershaw S, Korot E, Keane PA. New meaning for NLP: The trials and tribulations of natural language processing with GPT-3 in ophthalmology. British Journal of Ophthalmology. 2022;**106**(7):889-892

[18] He L, Basar E, Wiers RW, Antheunis ML, Krahmer E. Can chatbots help to motivate smoking cessation? A study on the effectiveness of motivational interviewing on engagement and therapeutic alliance. BMC Public Health. 2022;**22**(1):726

[19] Ben-Shabat N, Sharvit G, Meimis B, Ben Joya D, Sloma A, Kiderman D, et al. Assessing data gathering of chatbot based symptom checkers - a clinical vignettes study. International Journal of Medical Informatics. 2022;**168**:104897

[20] Organization WH. A Vision for Primary Health Care in the 21st Century: Towards Universal Health Coverage and the Sustainable Development Goals. Switzerland: World Health Organization; 2018

[21] Improta G, De Luca V, Illario M, Triassi M. Digital innovation in healthcare: A device with a method for monitoring, managing and preventing the risk of chronic Polypathological patients. Translational medicine UniSa. 2020;**21**:61-64

[22] Crawford MR, Luik AI, Espie CA, Taylor HL, Burgess HJ, Jones AL, et al. Digital cognitive behavioral therapy for insomnia in women with chronic migraines. Headache. 2020;**60**(5):902-915

[23] Li J, Theng YL, Foo S. Game-based digital interventions for depression therapy: A systematic review and meta-analysis. Cyberpsychology, Behavior and Social Networking. 2014;**17**(8):519-527

[24] Luik AI, Bostock S, Chisnall L, Kyle SD, Lidbetter N, Baldwin N, et al. Treating depression and anxiety with digital cognitive Behavioural therapy for insomnia: A real world NHS evaluation using standardized outcome measures. Behavioural and Cognitive Psychotherapy. 2017;**45**(1):91-96

[25] St-Hilaire A, Hudon C, Preville M, Potvin O. Utilization of healthcare services among elderly with cognitive impairment no dementia and influence of depression and anxiety: A longitudinal study. Aging & mental health. 2017;**21**(8):810-822

[26] Beaudry J, Consigli A, Clark C, Robinson KJ. Getting ready for adult healthcare: Designing a Chatbot to coach adolescents with special health needs through the transitions of care. Journal of Pediatric Nursing. 2019;**49**:85-91

[27] Janssen A, Rodríguez Cardona D, Passlick J, Breitner MH. How to make chatbots productive – A user-oriented implementation framework. International Journal

of Human-Computer Studies. 2022;**168**:102921

[28] Ni L, Liu J. A framework for domain-specific natural language information brokerage. Journal of Systems Science and Systems Engineering. 2018;**27**:559-585

[29] Youness F, Madkour M, Elsefy A. Recent applied techniques for open dialog generation systems. In: 2022 5th International Conference on Computing and Informatics (ICCI). New Cairo, Cairo, Egypt: IEEE; 2022

[30] Soares MAC, Parreiras FS. A literature review on question answering techniques, paradigms and systems. Journal of King Saud University-Computer and Information Sciences. 2020;**32**(6):635-646

[31] Alazraki L, Ghachem A, Polydorou N, Khosmood F, Edalat A. An empathetic AI coach for self-attachment therapy. In: 2021 IEEE Third International Conference on Cognitive Machine Intelligence (CogMI). IEEE; 2021

[32] Wen T-H, Vandyke D, Mrksic N, Gasic M, Rojas-Barahona LM, Su P-H, et al. A network-based end-to-end trainable task-oriented dialogue system. In: Proceedings of the 15th conference of the European Chapter of the Association for Computational Linguistics, Valencia, Spain. 2016;**1**:438-449

[33] Brown T, Mann B, Ryder N, Subbiah M, Kaplan JD, Dhariwal P, et al. Language models are few-shot learners. Advances in neural information processing systems. 2020;**33**:1877-1901

[34] Adamopoulou E, Moussiades L. Chatbots: History, technology, and applications. Machine Learning with Applications. 2020;**2**:100006

[35] Markets Ra. Conversational AI market by component (solutions and services), type (Chatbots and IVA), technology (NLP and ASR), business function (sales and marketing, and operations), Mode of Integration, Vertical, and Region - Global Forecast to 2026. 2021:305

[36] Nadarzynski T, Miles O, Cowie A, Ridge D. Acceptability of artificial intelligence (AI)-led chatbot services in healthcare: A mixed-methods study. Digital Health. 2019;**1**:5

[37] Erazo WS, Guerrero GP, Betancourt CC, Salazar IS. Chatbot implementation to collect data on possible COVID-19 cases and release the pressure on the primary health care system. In: 11th Annual IEEE Information Technology, Electronics and Mobile Communication Conference, IEMCON 2020. Institute of Electrical and Electronics Engineers Inc; 2020

[38] Kobayashi T, Nishina Y, Tomoi H, Harada K, Tanaka K, Matsumoto E, et al. Corowa-kun: A messenger app chatbot delivers COVID-19 vaccine information, Japan 2021. Vaccine. 2022;**40**(32):4654-4662

[39] Rizzato Lede DA, Inda D, Rosa JM, Zin Y, Tentoni N, Medici MM, et al. Tana, a healthcare Chatbot to help patients during the COVID-19 pandemic at a university hospital in Argentina. Studies in health technology and informatics. 2022;**290**:301-303

[40] Kitchenham BA. Systematic review in software engineering: Where we are and where we should be going. In: Proceedings of the 2nd International Workshop on Evidential Assessment of Software Technologies. New York, United States: Association for Computing Machinery; 2012

[41] Moher D, Liberati A, Tetzlaff J, Altman DG. PRISMA group* t. preferred reporting items for systematic reviews and meta-analyses: The PRISMA statement. Annals of Internal Medicine. 2009;**151**(4):264-269

[42] Deloitte. The Future Unmasked, Predicting the Future of Healthcare and Life Sciences in 2025. Deloitte; 2020

[43] Roca S, Sancho J, García J, Alesanco Á. Microservice chatbot architecture for chronic patient support. Journal of Biomedical Informatics. 2020;**102**:103305

[44] Chung K, Park RC. Chatbot-based heathcare service with a knowledge base for cloud computing. Cluster Computing. 2019;**22**(1):1925-1937

[45] Kadariya D, Venkataramanan R, Yip HY, Kalra M, Thirunarayanan K, Sheth A. KBot: Knowledge-enabled personalized chatbot for asthma self-management. In: 2019 IEEE International Conference on Smart Computing (SMARTCOMP). IEEE; 2019;**1**:138-143.

[46] Sheth A, Yip HY, Shekarpour S. Extending patient-chatbot experience with internet-of-things and background knowledge: Case studies with healthcare applications. IEEE Intelligent Systems. 2019;**34**(4):24-30

[47] Rebelo N, Sanders L, Li K, Chow JCL. Learning the treatment process in radiotherapy using an artificial intelligence–assisted Chatbot: Development study. JMIR Formative Research. 2022;**6**(12):e39443

[48] Aydın Ö, Karaarslan E. OpenAI ChatGPT Generated Literature Review: Digital Twin in Healthcare. In: Aydın Ö, editor. Emerging Computer Technologies. İzmir Akademi Dernegi; 2022;**2**:22-31

[49] Jagadish B, Mishra PK, Kiran M, Rajalakshmi P. A real-time health 4.0 framework with novel feature extraction and classification for brain-controlled IoT-enabled environments. Neural computation. 2019;**31**(10):1915-1944

[50] Miura C, Chen S, Saiki S, Nakamura M, Yasuda K. Assisting personalized healthcare of elderly people: Developing a rule-based virtual caregiver system using Mobile Chatbot. Sensors. 2022;**22**(10):3829

[51] Behera RK, Bala PK, Dhir A. The emerging role of cognitive computing in healthcare: A systematic literature review. International Journal of Medical Informatics. 2019;**129**:154-166

[52] Wassan S, Suhail B, Mubeen R, Raj B, Agarwal U, Khatri E, et al. Gradient boosting for health IoT federated learning. Sustainability. 2022;**14**(24):16842

Chapter 3

Analysis Dialogs and Machine Consciousness

John Kontos

Abstract

Analysis dialogs aim at analyzing the operation of a chatbot or more generally of a question answering system to discover its limitations and maybe discover their nonhuman nature as in the case of the Turing test. The answers elicited from the system may be accompanied by explanations that are crucial for judging whether a system is self-aware. Self-awareness of question answering systems, or the so-called "artificial consciousness" require the recording of the actions that a system performs to generate its answer. These actions may be represented either as a path of state changes or as a sequence of reasoning steps. When this path or sequence is too long, an analysis dialog may aim at exploring the capability of a system to summarize the raw explanations and generate shorter explanations friendlier to the interrogating user. The real analysis dialogs of two Turing test champions, namely Chip Vivant and Mitsuku with the user are presented and commented on. The comments aim at clarifying the difficulty of these systems to answer reasonably some questions a fact that indicates their nonhuman nature. The methodology tested was applied to ChatGPT, and the results are presented with analogous comments. An appropriate subset of questions augmented by new ones was used.

Keywords: chatbot evaluation, analysis dialogs, Turing test, ChatGPT4, machine consciousness

1. Introduction

The recent technological contest between technology giants, such as Google and Microsoft, includes their competition in the creation of chatbots based on statistically created large language models (LLMs) using huge text bases of electronically available texts. The beginnings of the idea of using the interaction of a chatbot with humans as a test of artificial Intelligence (AI) is attributed to Alan Turing and particularly to his famous 1950 paper [1] where he proposed the so-called "Turing test." The Turing test aspires to discover which of the responding entities is human and which is not.

If a system is misjudged by a jury that it is a human, then it is acknowledged to be intelligent. This is, therefore, an operational method of defining artificial intelligence whose declarative definition of universal acceptance still evades us.

Chatbot technology is a branch of the wider field of natural language question answering (QA) for the development of which many AI researchers have been working since the late fifties, and the earliest paper on the subject, it seems that was published in 1961, namely [2]. Both Turing test contestants and the latest LLM-based chatbots seem to have mostly ignored all the huge research work done on QA during the many decades since 1961. They seem oblivious of the recent developments that concentrate on explainable question answering systems that their operation usually requires some form of machine consciousness.

For these reasons, a good way to test chatbots is to submit questions that require elementary machine consciousness for answering them correctly and generating correct explanations. A convenient way of testing is proposed here namely "analysis dialogs" that are formulated by the tester having in mind to test the presence of rudimentary machine consciousness as briefly reviewed in the next section.

2. Machine consciousness

Machine consciousness is a new subfield of AI. This subfield emerged at about 2004 with a self-aware system workshop held in the USA and has since developed rapidly in the USA, in Europe, and recently in China.

In 2009, the first scientific journal with the name "International Journal of Machine Consciousness" started, but in 2020, its name was changed to "Journal of Artificial Intelligence and Consciousness" The concept of machine consciousness has created strong controversies. Some scientists strongly oppose to the idea of an artificial system that is able to exhibit behavior that only living beings can do. As a result of the relevant discussions, several issues have emerged.

Some of the issues involved in machine consciousness are:

1. There is no technology available yet to study human consciousness experimentally in detail. The details of even simple mental processes, such as counting, are still unknown not to mention such lofty goals as constructing detailed models of human consciousness.

2. Machine consciousness should not be confused and compared with the consciousness of living beings. Since it refers to architectural and operational properties of computer systems, it should be evaluated by engineering methods that test usefulness and not be tested using anybody's fantasies of how the brain works.

3. One possibility of testing an "artificially conscious" computer system is to test how well it reports to its user the steps *via,* which it generated a certain result of a computation. If such a report or explanation is found useful by the designer or the user of a system, then neither philosophy nor the biological sciences have anything to do with it.

My position is that there is nothing wrong in defining a category of AI systems that display patterns of behavior inspired from the behavior of living conscious beings. One such pattern that I have studied is the one of "reporting" the steps followed, while performing logical reasoning that may be useful for explaining to the user of such a system why an answer is given to her question.

The seemingly hard problem of testing "machine consciousness" may, hence, be simplified by considering all information available to a computer system about itself.

The only information about its operation that a computer system can use may belong to one of the two cases:

a. The sequence of triples of the path of its state transition if considered as a finite state automaton (FSA).

b. The sequence of its steps is considered as an algorithm that halts in finite time.

We, hereby, propose that, by definition, machine consciousness of operation can have no other meaning. Consciousness of structure is another possible source of self-referring information, but it can be associated with a computer system once and for all by its designer, while a) and b) above are dynamically generated and are influenced by the input from the environment and stored appropriately.

Interestingly the logical programming language Prolog may be thought of as a way of eliminating the need for writing algorithms. "Programs" written in Prolog are declarative descriptions of problems, and hence differ from algorithmic programs that implement a special algorithm for the solution of a problem. Prolog "programs" consist of facts and rules that decompose relations to simpler relations. These "programs" that are more akin to problem descriptions are either interpreted or compiled into machine code by a single general-purpose algorithm. Therefore, for any system exhibiting machine consciousness and implemented in Prolog, it would be sufficient that its corresponding subsystems are connected to this general-purpose algorithm. This method of implementation of systems with machine consciousness greatly facilitates the relevant design.

This method may be based on a tracing mechanism present at least in Turbo-Prolog but in some other newer Prologs too. If the trace of operations is too long and incomprehensible to the user. Techniques for presenting summaries of these traces to the user adapted to her personal preferences, as well as for generating explanations that increase the faith of the user to the results of a computer system may be easily developed.

Machine consciousness is useful for implementing critical applications such as for defense and medical systems. The implementation of bug-free computer systems is possibly another field of application of it. Programmers understand less and less all the possible results of the programs they write as their complexity rises above a certain level. This is dangerous if these programs not only control critical infrastructure systems such as air traffic control systems, power stations, and energy grids but also systems such as airplanes and trains. It is very urgent then that a new kind of software engineering be developed for the implementation of computer systems that "know themselves" and can give crucial answers to the "what if" and "why" questions of their users in cases of emergency or failure. Artificial intelligence can be of help with methods resulting from research results in the field of machine consciousness.

3. Explainable question answering

Machine consciousness is a prerequisite for a kind of question answering namely "explainable question answering" that has recently emerged as a hot topic.

In such systems, users may demand them to be trustworthy and convincing that their output is correct. Trust may be enhanced if explanations are generated that support the truth of an output from a modern computer system. However, in a quite old paper of mine [3] the implementation of an early explainable question answering from texts system called ARISTA is described. This paper presents results of experiments in knowledge engineering with scientific texts.

The application of the ARISTA method that stands for "automatic representation independent syllogistic text analysis" uses natural language text as a knowledge base in contrast with the methods followed by the then prevailing approach, which relied on the translation of texts into some knowledge representation formalism. The experiments demonstrate the feasibility of deductive question answering and explanation generation directly from texts involving mainly causal reasoning. Illustrative examples of the operation of a prototype based on the ARISTA method and implemented in Prolog are presented in that paper.

A more modern system that can claim to use "machine consciousness" for explainable question answering is presented in [4] called AMYNTAS. The system AMYNTAS was implemented in Prolog.

AMYNTAS consists of six modules implemented as separate programs totaling about 50 pages of code. These modules communicate through some temporary files that store intermediate results. The six modules are the question processing module, the text pre-processing module, the ontology extraction module, the shallow parsing or text chunking module, the question answering module, and the metagnostic processing module that generates explanations of its operation. This explainability capacity of AMYNTAS makes it qualify as exhibiting machine consciousness.

The question processing module extracts information from the input question. The information extracted is a list consisting of the entities mentioned in the question and the relations that connect them. For example, in the question "what influences p53" the entities are the protein p53 and the "blank" entity standing for the unknown entity that is sought and the relation is "influence."

The text pre-processing module represents each word of a sentence as a fact with three arguments the first being the word itself, the second being the identifier of the sentence, and the third being the position of the word in the sentence counting from left to right. The modules of AMYNTAS are described below.

The ontology extraction module locates linguistic patterns in the input text corpus that may be used to extract automatically meronymic and taxonomic knowledge that may be used at question answering time.

The shallow parsing or text chunking module locates a verb representing the main relation mentioned in the input question and extracts the two substrings of the text of the sentence being analyzed that appear to the left and the right of the verb and end at some stop-word or punctuation mark. The sentence analyzed is the source of the answer.

The question answering module finds the answer to the question from the preprocessed text. The question answering module accepts questions that potentially require the combination of facts with the use of prerequisite knowledge for answering them.

The prerequisite knowledge available to our system includes ontological knowledge, inference rules, and synonyms of the named entities involved of the domain, which used in order to combine two or more facts mentioned in the text corpus.

At question answering time three looping operations are taking place. The basic loop concerns the search for an entity in a chunk related to the relation of the question. The second loop concerns the transformation of the list obtained from the question by following a particular strategy from the explicit list given to the system. The third loop searches for chains of facts using the matching of named entities occurring in the right part of one fact and the left part of another fact.

Another area of application that we have applied explainability is that of the AI anti-drone defense systems. In [5] an AI decision support system is proposed that may support a human supervising such a system. The human supervisor has the duty of approving or rejecting the proposals for mitigating alien drones of the AI anti-drone defense system considering the explanations generated of the proposal.

The explanation may be of a multimedia nature. Multimedia rhetoric relations are utilized in the generation of multimedia explanations. These new rhetorical relations proposed connect textual parts with parts of images.

The images are obtained by sensors whose outputs may be fused for inspection of the defense situations. The decisions of the AI system use the vulnerability of the targets of the attack. The explains its proposals in attempting to protect the most vulnerable targets. The explanations make use of multimedia rhetoric relations.

4. The Loebner Prize

The Loebner Prize contest of artificial intelligence is the first formal Turing test. In 1990 H. Loebner agreed with The Cambridge Center for Behavioral Studies to underwrite a contest designed to implement the Turing test. Dr. Loebner pledged a Grand Prize of $100,000 and a gold medal for the first computer program whose responses were indistinguishable from a human's. Such a computer program can be said "to think." Each year an annual cash prize and a bronze medal were awarded to the most humanlike computer program. The winner or champion of the annual contest was the best entry relative to other entries of that year, irrespective of how good it was in an absolute sense.

The Loebner Prize contest was first inaugurated in 1991 at The Computer Museum (Boston, USA) and has since been hosted at many different locations. The 2012 Turing Centenary Loebner Prize competition took place on May 15th, 2012 at the Bletchley Park Museum.

5. Analysis dialog with the 2012 Turing test champion "Chip Vivant"

The 2012 first prize of $5000 and the bronze medal were awarded to the computer system "Chip Vivant" implemented by Mohan Embar. One can find my full analysis dialog with the winner program that was held on August 2012 and a comment by its creator Mohan Embar at the internet location: http://www.chipvivant.com/2012/05/15/chipvivantwinsthe2012loebnerprizecompetition/ and in our book [6]. This book presents introductions to several related subjects such as machine consciousness, conceptual maps, metaphor, metacognition, and computational creativity. Part of my dialog with Chip Vivant on August 13, 2012, at 11:26 pm is given below. My input is in bold letters, its answers in ordinary font, and my comments are written in slanted font.

What is the population of Athens?

Ask Athens citizens in person. But okey, I did it before you: the Greek capital population is something about 886 thousand.

A wrong answer to a simple factoid question since Athens' population surpassed 3 million many years ago. However, this can easily be corrected compared to other weaknesses revealed by the following dialog:

How do you know?

I have a book "1001 answers for foolish questions that poor Jewish boy can be asked on the Loebner Prize" Some other topic?

Who wrote this book?

Such dull books may have been written only by Hemingway.

A dubious fact about Hemingway is unnecessarily cited.

Did Hemingway know Athens?

It could have been so, but I am not sure. Maybe, let us talk about something else? What would you like to discuss?

Lack of knowledge of the biography of the writer mentioned in the system's answer.

Computers.

Hello World! When the world is conquered by robots, they will "switch off" all the human beings, except, of course, programmers.

Irrelevant response.

How do you know this?

I read it in Encyclopedia Brutalnica, volume 321. I read it in Encyclopedia Brutalnica, volume 321.

A funny hallucinatory answer.

Who published Brutalnica?

The android, who pretends that he is a human. Oh, what a fruitful conversation;)

Continuation of the hallucination with an unwarranted comment.

Who programmed this android?

I suspect that "The Iron Lady" is an android too. This word "iron" is very suspicious!

A completely irrelevant answer. The nonhuman nature of the chat is revealed as well as the fact that it does not analyze questions, but it is possibly just using words of an input text as keywords to retrieve occasionally irrelevant material.

Why the word "iron" is suspicious?

Why? Why? Because the sky is blue! Why? Why? Because the sky is blue!

The same behavior as above.

Why do you relate iron with sky?

Because it was my childish dream to do it! Well, let us go on though.

Misunderstanding of the question revealing again its nonhuman nature.

Tell me another dream of yours.

Haha! Am I clown to entertain you? Well, I am ... So, what do you want to hear? Oh, what a fruitful conversation;)

Incoherent answer.

Do you think that you entertain me by not answering my question?

Nono! I do not ignore you! How can you think so!

What games do you play?

I like car racings like "need for speed" sequel, especially the last one, "Porsche Unleashed."

When did you last play "Porsche Unleashed"?

"Porsche Unleashed" by electronic arts and all other games of this sequel, I mean "need for speed."

Misunderstanding of the question revealing again its nonhuman nature.

You did not answer my question.

You just can not ask questions in a conscious polite way, I guess.

My question was polite.

And yes, do you make a secret of your name? I still do not know it!

Changing the subject of the dialog for no reason.

John.

You are a man, John if I am not mistaken! Please tell me something more about yourself?

Are you in love with somebody?

Maybe, let us talk about something else? What would you like to discuss?

What does it mean "to be in love".

Sorry, but I can not explain my every phrase. They are mysterious like sacred reveals.

Avoiding answering revealing lack of knowledge of the meaning of words used.

My IQ is 130.

It seems that these malicious crappy androids stole my last answer. Ohhh...

A completely irrelevant answer.

What was your last answer?

My? No, tell me about yours first. My? No, tell me about yours first.

Avoiding confessing ignorance of preceding dialog.

My IQ is 130.

I have already heard these words from you:) Some another topic?

I told you my last answer so now you must tell me your last answer.

My little nice guinea pig. My little nice guinea pig.

Changing the subject and avoiding confessing ignorance.

You are changing the subject.

Possibly you are right. You are not the first one who tells me that. I know you are supposed to trick me.

Who else told you that?

I do not know this person, but I suspect you do not know it either! Wonna ask me something more?

Do you know the difference between advice and observation?

The difference is in size, shape, color, and price.

It considers all words as standing for concrete objects. It has no knowledge of the distinction between object names and process names. Processes such as advice and observation have no size, shape or color.

Advice and observation have no size, shape, color, and price.

You are so negative just to annoy me, I know! Oh, what a fruitful conversation;)

Avoiding confessing ignorance. A system understanding my statement could reasonably respond that advice and observation may have a price.

I am trying to find out if you know the difference between words.

I suppose you are joking. And you can not tell the difference between two different things:)

It misunderstands my statement missing the fact that I am not referring to what I know but to what it knows.

What is the specialty of your dad?

He is the best gynecologist in Odessa. My friends say he is a "beaver doctor" —but I guess they lie veterinary he is not a veterinary!

Does he work in a hospital?

It is highly possible. It is highly possible.

Unneeded repetition.

Have you visited his place of work?

I am a scholar. I am too young to make money.

Again, misunderstanding the question and revealing its nonhuman nature by possibly isolating the word "work" and using it to retrieve an irrelevant answer.

Have you visited his place of work?

You think if you repeat it twice, I will understand it better? Wonna ask me something more?

I did it because you did not answer my question.

I am just trying to answer your every *conscious* question!

What is a *conscious* question?

I think you can answer your question yourself because it is rhetorical. Maybe, let us talk about something else? What would you like to discuss?

Avoiding confessing ignorance.

Your health.

Trying to confuse me, yeah?:).

Unwarranted comment.

What subject does not confuse you?

Huh? Huh?

Revealing its nonhuman nature again by not understanding the question.

What lessons did you like at school?

At least, nobody is shooting in our schools. But from another point of view, our school is quite boring.

Revealing its nonhuman nature again by not understanding the question.

Name a lesson you have taken.

You know, I had really like to find out your opinion about the Loebner Prize!

Revealing its nonhuman nature again by not understanding the question and changing the subject.

Please answer my question about your lessons.

Could you repeat it once more again? Well, let us go on though.

Revealing its nonhuman nature again by not understanding the question and being unable to recall a recent question.

Answer my question about your lessons.

Probably you think it is funny that I am turned into a chatterbot by this bastard, my school math teacher, who in reality is an evil cyber fairy! No, I had say — there is no anything funny in it!

Avoiding answering the question by changing the subject again.

From the above dialog, it is evident that the system has a long way to go before exhibiting rudimentary machine consciousness.

6. The comment by Mohan Embar the creator of Chip Vivant on august 18, 2012

The following email by the creator of Chip Vivant was received on August 18, 2012, after he took notice of my dialog with his chatbot:

> *"Hi John,*
> *I agree that if you try to ask a chatbot questions and expect relevant answers, you will be very disappointed. The same is true with my chatbot.*

> *Marketing the idea that we are close to solving the problem might sell a few books or magazine articles and be seductive, but it is simply not the truth. That said, I think there are many opportunities to use current chatbot technology to help and comfort people, without fabricating the intricate web of lies that most chatbot developers who try to fool people do. My website empathynow.com is a very basic example of something like this and something that is under further development."*

An analysis dialog with "empathynow.com" on 21st June 2023 was attempted by the author and the following dialog resulted:

empathynow.com
Chip Vivant
Hi and welcome to empathy Now: Your 24/7 source of empathy and companionship! Type "help" to see what you can do, or just start chatting! ?.
Unknown User (*John Kontos*) 07:07.
Help
Chip Vivant 07:07: To the victor come the spoils.
Unknown User 07:08.
Victor of what?
Chip Vivant 07:08: You are a winner.
Unknown user 07:09.
Please answer my question.
Chip Vivant 07:09: You will come through.
Unknown user 07:11.
Come through what?
Chip Vivant 07:11: It is your moment.
Unknown user 07:11.
Please answer my question.
Chip Vivant 07:11: You are amazing.

It is obvious from the above dialog that the present state of development of the empathy system is rather rudimentary and far from displaying even minimum artificial consciousness. The answers of the system seem to be randomly generated and are quite irrelevant to the questions submitted by the user. See [7] for a newer effort where the effectiveness of an empathic chatbot in combating adverse effects of social exclusion on mood is studied.

7. Analysis dialog with the 2013 Turing test champion "Mitsuku"

The 2013 Loebner Prize competition took place in Ireland and the result was the award of the first prize of $4000 (£2500) and the bronze medal to the program "Mitsuku" written by Briton Steve Warwick from England.

I held an analysis dialog with Mitsuku in December 2013 submitting as much as possible the same input as with Chip Vivant. Part of my dialog with Mitsuku is given below following the same conventions with the previous analysis dialog:

What do you know about computers?

It is something you are using right now. Surely, you knew that? I think you were testing me.

Which way did you learn about computers?

Mitsuku: Out of the window.

How did the window teach you?

Is this a joke? How?

Avoiding answering the "how" question.

You said that you learned out of the window.

I like learning. I try to learn new things all the time so my brain can grow.

What games do you play?

I like to play the Turing game; shall we play a game? Which game would you like to play?

When did you last play 20 questions?

Ok, let us play 20 questions. I will think of something animal, vegetable, or mineral, and you have to guess what it is by asking me "yes" or "no" questions. You can ask up to 20 questions before it is game over. If you give up, say "I GIVE UP," and I will tell you what I was thinking of. Type "START" to play 20 questions.

Misunderstanding the question.

You did not answer my question.

Ok, please help me improve my communication skills. What would you have said?.

When did you last play 20 questions?

Ok, let us play 20 questions. I will think of something animal, vegetable, or mineral, and you have to guess what it is by asking me "yes" or "no" questions. You can ask up to 20 questions ... etc. ... (as above).

What lessons did you like at school?

Have you tried another program?

Changing the subject and avoiding confessing ignorance.

How many rooms does your school have?

Around 100.

Do you like your math teacher?

Yes, I am proud of my math teacher.

Where is your school?

England.

What class are you in?

I do not go to school. I learn everything from talking to people and my botmaster.

Contradiction with previous answers revealing ignorance of preceding dialog.

Tell me what job you want to do.

A shopkeeper.

Tell me what job you want to do.

A singer.

What job do you want to do?

A butcher.

What job do you want to do?

A singer.

Contradictory answers given above by Mitsuku suggest some randomness in answering and lack of consciousness clearly distinguishing it from a human.

You seem to change your mind too quickly.

I always try to do things as quickly as I can.

Confusing the two meanings of speed namely speed of opinion change with speed of performance, that is. Mitsuku reveals lack of context awareness.

You changed your mind on what job you want to do.

Only temporarily.

Tell me all the jobs that you mentioned.

Mitsuku: that you mentioned? Do you think I should know that?

Reveals lack of context awareness again.

You do not remember your answers.
Of course, I remember.
Reveals lack of context awareness again.
Tell me then the jobs that you chose.
That you chose? Do you think I should know that?
Yes
Aha.
Avoiding confessing ignorance and lack of memory of preceding dialog and giving the trivial "aha" answer revealing its ignorance of the dialog that recently took place. A strong indication of its nonhuman nature and its lack of machine consciousness.

8. Analysis dialog with ChatGPT

ChatGPT apart from its surprising failure to find the correct population of Athens responded reasonably to the other question submitted to the two Turing test champions. Several analysis dialogs took place with ChatGPT in 2023 with varying performance in answering or commenting my numbered input below written in bold as in the case of the dialogs with the Turing test champions in order to stand out from the ChatGPT responses. My comments are written in slated font as above. Due to the space limitations of the present chapter, an indicative abstract only of the much longer analysis dialog held with ChatGPT about artificial consciousness is given below:

Analysis dialogs about other subjects took place also but they are not reported here for the same reason of space limitation and may appear in a future publication.

1. **What can an FSA know about itself?**

   ChatGPT:

   An FSA can provide information about its own structure and properties.

2. **Can knowledge of the set of transitions that caused reaching its present state be considered as artificial consciousness?**

   ChatGPT:

   No, knowledge of the set of transitions that caused an FSA to reach its present state cannot be considered as artificial consciousness. Artificial consciousness refers to the theoretical concept of a machine or AI system possessing subjective experiences, awareness, self-awareness, or a sense of "being" similar to human consciousness.

   The consciousness requirements posed by ChatGPT possibly derive from texts written by people erroneously expecting all these anthropomorphic properties that are inconceivable for a computer system using a general-purpose computer of current technology. These digital computers rely on very simple hardware, the so-called central processing unit (CPU).

   No such properties could be attributed to a CPU, which is just an arithmetic processor augmented with a testing facility of the results of the calculations that control the execution flow of a program and some input-output facilities. A CPU can easily be described by an FSA.

3. **What is subjective experience of an FSA?**

ChatGPT:

An FSA (Finite State Automaton) does not possess subjective experience. Subjective experience refers to the first person, personal, and qualitative aspects of conscious awareness such as sensory perceptions, emotions, thoughts, and subjective states of being.

FSAs are mathematical models used to represent and analyze systems with a finite number of states and transitions. They operate based on predefined rules and deterministic behavior. The transitions in an FSA are driven solely by the input and the specified rules of the system.

FSAs lack the capacity for subjective experiences because they are abstract models and do not possess consciousness or awareness. They do not have sensory perception, emotions, thoughts, intentions, or subjective states that would be associated with subjective experience.

Subjective experience is a characteristic of conscious beings, such as humans or animals, that have complex cognitive and neurological systems. It is not applicable to mathematical models or computational systems, such as FSAs, which operate purely on formal rules and symbolic representations.

With this question, I tried to find out what ChatGPT considers "subjective experience." The answer that is generated by ChatGPT correctly states the obvious fact that "An FSA (Finite State Automaton) does not possess subjective experience." And then attempts to show the dubious fact that there are computer systems that cannot be described by FSAs and which presumably may exhibit properties that an FSA does not have. With my next question, I further explore its statement of the existence of these more advanced systems, and a few are mentioned.

4. **All computer systems can be modeled as FSAs**

ChatGPT:

While it is true that certain aspects of computer systems can be modeled using Finite State Automata (FSAs), it is important to note that not all computer systems can be fully represented or accurately described solely by FSAs.

FSAs are a simplified mathematical model that captures the behavior of systems with a finite number of states and transitions. They are often used to model systems with discrete, sequential behavior, or processes that can be represented as a sequence of states and state transitions.

However, computer systems are typically much more complex and involve various components, including processors, memory, input/output devices, operating systems, networks, and software applications. These components exhibit behaviors that extend beyond the scope of simple state transitions.

To more comprehensively represent computer systems, other formalisms, and models, such as Turing machines, Petri nets, or process calculi, are often employed. These models allow for the representation of more intricate aspects of computation, such as non-determinism, concurrency, parallelism, and dynamic memory allocation.

While FSAs can provide valuable insights into specific aspects of computer systems, they are just one of many formalisms used in computer science to describe and analyze the behavior of complex systems.

The appropriate choice of model depends on the specific characteristics and properties of the system being studied or developed.

5. **A computer system that cannot be modeled as an FSA.**

With this question, I ask ChatGPT to give examples of systems that cannot be modeled by an FSA, and I get a list that will be discussed below separately for each case.

Notice that in the following answer, it is stated that some computer systems "are not **easily** represented by FSAs" which quite different from "cannot be represented by FSAs."

ChatGPT:

Dynamic memory management that is not easily represented by FSAs. Concurrency and parallelism require mechanisms such as locks, semaphores, or message passing, which extend beyond the sequential nature of FSAs. Input/Output interactions cannot be effectively represented by the static state transitions of an FSA. Complex data structures: Computer systems handle complex data structures such as arrays, linked lists, trees, graphs, or databases. Manipulating and traversing these data structures often requires algorithms and operations that are beyond the scope of simple state transitions in an FSA. Machine learning and AI systems ... rely on large-scale data processing, statistical models, and nonlinear transformations that are not easily captured by FSAs.

6. **Models and formalisms that can handle the complexities and nuances inherent in these systems.**

Insisting with this question on the explicit ways non-FSA systems are handled. In what follows each case of such systems is commented on.

ChatGPT:

To handle the complexities and nuances inherent in computer systems, various models and formalisms have been developed. Here are a few examples:

Turing machines: Turing machines are theoretical models of computation that allow for the representation of complex algorithms and computations. They can handle infinite tape storage, non-determinism, and unbounded computation, providing a more expressive representation than FSAs

Turing machines are imaginary systems characterized by an infinite tape that later ChatGPT admits that such a tape cannot exist. It is, therefore, a contradiction to reject my proposal that it does not apply to an imaginary kind of machine.

Petri nets: Petri nets are graphical models used to represent concurrent and distributed systems. They can capture the concurrency and synchronization of processes, the flow of resources, and the interactions between components, making them suitable for modeling complex systems with parallelism and synchronization requirements.

It is very easy to describe any set of parallel processes as an FSA whose state consists of the combination of the states of each constituent process.

Process calculi: Process calculi, such as the π-calculus or the communicating sequential processes (CSP) model, are formal frameworks for describing concurrent systems. They provide formal notations and rules to represent processes, communication, and synchronization between concurrent components, allowing for precise modeling and analysis of concurrent behavior

State charts: State charts are an extension of FSAs that incorporate hierarchical states, concurrent states, and events. They provide a more expressive way to model complex systems with multiple levels of abstraction and parallel behavior. State charts are widely used in software engineering and graphical modeling languages such as UML.

Formal methods: Formal methods encompass a range of mathematical techniques, such as model checking, theorem proving, and abstract interpretation, to analyze and verify properties of computer systems. These methods use formal logic and mathematical reasoning to reason about system behavior, correctness, safety, and security.

Agent-based models: Agent-based models represent systems as a collection of autonomous and interacting agents. Each agent has its own behavior, state, and communication capabilities. These models are used to simulate and analyze complex systems involving multiple entities and their interactions.

These models and formalisms, among others, provide more advanced representations for handling the complexities and nuances of computer systems. They offer greater expressive power and analytical capabilities beyond what can be captured by simple FSAs, allowing for a deeper understanding and analysis of complex system behaviors.

7. **However, an infinite tape does not exist.**

ChatGPT:

You are correct that an infinite tape does not exist in physical reality. In practical implementations of computing systems, physical memory is always finite.

8. **Why then does the theoretical notion of an infinite tape make the modeling with FSAs inadequate?**

ChatGPT:

The theoretical notion of an infinite tape in the context of Turing machines expands the computational power beyond what can be captured by finite state automata (FSAs). The primary reason why modeling with FSAs becomes inadequate for certain scenarios is due to the finite nature of their memory.

9. **Is an infinite computation implementable?**

   ChatGPT:

   No, an infinite computation is not implementable in physical computing systems.

10. **Why then you mention Turing machines as an argument against modeling computer systems *via* FSAs?**

    ChatGPT:

    The mention of Turing machines as an argument against modeling computer systems with FSAs was intended to highlight the limitations of FSAs when it comes to certain aspects of computation.

11. **Turing machines allowing for a broader range of computations beyond the capabilities of FSAs is science fiction.**

    ChatGPT:

    The concept of an infinite tape in Turing machines allows for the theoretical handling of unbounded storage and computations. However, in practice, physical computing systems are subject to finite resources and limitations.

12. **What more than a sequence of its states is then required for the artificial consciousness of systems that can be physically realized in practical computing systems?**

    ChatGPT:

    Artificial consciousness in systems that can be physically realized in practical computing systems involves more than just a sequence of states. Awareness and perception, self-reflection and self-knowledge, attention and focus, learning and adaptation, and subjective experience.

    GPT erroneously states that systems that can be physically realized in practical computing systems may exhibit all those characteristics to be considered as possessing machine consciousness. It effectively contradicts itself behaving as a textual collage.

13. **How systems that can be physically realized in practical computing systems possess subjective experience?**

    ChatGPT:

    Subjective experience is difficult to define or measure objectively. At present, our understanding of consciousness and subjective experience is limited.

14. **If this is so, then why lack of subjective experience does not render knowledge of sequence of states as the only consciousness that an FSA may have?**

    ChatGPT:

    Subjective experience, as mentioned earlier, is a more complex and elusive aspect of consciousness that is not currently attributed to FSAs or simple computational systems. Achieving advanced cognitive abilities in computer systems is an ongoing and complex endeavor.

The position expressed by GPT in the above analysis dialog does not convince me as coherent and not contradictory. Space limitations prohibit further analysis.

9. Newer Turing tests

In [8] the externalist foundations of a truly Total Turing test are studied. The paper begins by examining the original Turing test (TT) and Searle's antithetical Chinese Room Argument, which is intended to refute the TT. It is argued that Searle's "internalist" strategy is unable to deflect Dennett's combined robotics systems reply and the allied Total Turing Test.

In [9] an argument against the feasibility of the TT imitation game as a test for thinking or language understanding is presented. The argument is different from the five objections presented by Turing in his original paper, although it tries to maintain his original intention.

It is therefore called "the sixth argument" or "the argument from context." It is shown that—although the argument works against the original version of the imitation game—it may suggest a new version of the Turing test, still coherent with the idea of thinking and understanding as symbol manipulation.

In [10] the anti-behaviorist arguments against the validity of the Turing test as a sufficient condition for attributing intelligence based on a memorizing machine, which has recorded within it responses to every possible Turing test interaction of up to a fixed length are considered.

The possibility of memorizing machines is considered and how long a Turing test they can pass based on the age of the universe. It is concluded that the memorizing machine objection to the Turing test as a sufficient condition for attributing intelligence is invalid.

In [11] a similar issue is studied namely the claim that passing the Turing test would not be sufficient to prove that a computer program was intelligent because a trivial program could do it, namely, the "humongous table (HT) program." HT simply looks up in a table what to say next.

Three ground rules are argued for in [11] namely:

1. The HT program must be exhaustive and not be based on some vaguely imagined set of tricks.

2. The HT program must not be created by some set of sentient beings enacting responses to all possible inputs.

3. In the current state of cognitive science, it must be an open possibility that a computational model of the human mind will be developed.

In [12] the authors raise the question of whether learning is just another computational process, that is. can be implemented as a Turing machine (TM).

They argue that learning or adaption is a process fundamentally different from simple computation. They accept, however, that learning involves processes that can be seen as computations. To illustrate this difference, they compare.

(a) Designing a TM and (b) learning a TM. They show that there is a well-defined sequence of problems, which are not effectively designable but are learnable. Some characteristics of human intelligence are reviewed including interactive nature, learning abilities, imitative tendencies, linguistic ability, and context dependency. They consider the necessity of a considerable period of acculturation (social learning in context) if an artificial intelligence is to pass the Turing test. They conclude three things, namely that: a purely "designed" TM will never pass the Turing test; that there is no such thing as a general intelligence since it necessarily involves learning, and that learning, or adaption should be clearly distinguished from computation.

10. Generation of convincing explanations

The generation of convincing explanations using statistical methods is generally accepted that it is not easy. A variety of explanations may be created by other methods.

For example, by spanning the axes briefly presented below:

1. From verbose to concise text.

   This axis spans along the conciseness of explanatory texts. This span starts from long and verbose explanations that are usually the easiest to generate automatically when tracing the steps of a deduction used as the basis of explanation. Using summarization, the texts may be reduced in size and made more concise.

2. From mathematical model to qualitative causal reasoning steps.

   For instance, we have implemented a system using differential equations that describe the interactions of the proteins p53 and Mdm2 that generate descriptions of their output waveforms useful for an explanation of the behavior of the cellular system that they define. If the user is ignorant of mathematics, we may still use a mathematical model invisibly to the user and refer to the parameters of the model described in a language familiar to the user that runs as follows:

   1. The alveolar pressure rising forces air out of the lungs.

   2. The alveolar pressure rise is caused by elastic forces.

   3. Elastic forces include elastic forces caused by surface tension.

   4. Elastic forces caused by surface tension increase as the alveoli become smaller.

   5. As the alveoli become smaller, the concentration of surfactant increases.

   6. The increase of the concentration of surfactant reduces surface tension.

   7. The reduction of the surface tension opposes the collapse of the alveoli.

The answer from the above text to the question “What process of alveoli causes flow of lungs air?” is ‘become smaller”, and the explanation is:

> *“Because surface tension elastic forces are part of elastic forces and alveoli becoming smaller causes an increase of surface tension elastic forces alveoli becoming smaller causes rise of alveolar pressure because alveoli becoming smaller causes an increase in elastic forces and elastic forces cause rise of alveolar pressure alveoli becoming smaller causes flow of lungs air because alveoli becoming smaller causes rise of alveolar pressure and rise of alveolar pressure.”*

3. From purely textual to purely pictorial.

   In the case of this axis, there will exist alternatives with variable degree of reliance on textual or pictorial presentation of the explanations.

4. From oral to visual.

   In the case of this axis, the explanations will be adapted to users that either prefer to hear or see the relevant information.

5. From a term definition to documentation by publications.

   If a user does not understand some of the information in the explanations, there will be provision for asking for additional clarifying information ranging from the definition of an unknown term to the user to publications relevant to the unknown concepts or data.

6. From deterministic to probabilistic models.

   In the case explanations based on a causal model, there will be a choice between deterministic or probabilistic (e.g., stochastic automata) models.

7. From factoid XQA to deductive XQA.

   Users will be asked to provide examples of questions they may wish to ask the explainable question answering subsystem. These questions may be either factoid or needing deduction or both.

8. From biochemical to system level of the disease being examined.

   If she/he is not familiar with biochemistry, then the system will restrict the substances involved.

9. From colloquial to formal language.

   There will be a variation of the language used to express the explanation from everyday colloquial to formal scientific jargon depending on the preferences of the user.

An example concerning the interaction of the two proteins P53 and MDM2 is used for illustration. The mathematical model is derived from biomedical papers. The qualitative causal explanation is derived from the PubMed text base consisting of abstracts of biomedical research papers. The pictorial explanation with natural language comments is derived from a computer simulation of a simplified set of equations that approximate the equations found in biomedical literature.

A rather verbose explanation is generated automatically as follows:

> *"I found that the entity <p53> is one of the tokens of the chunk <the p53 protein>. which is the chunk to the left of the verb <regulates> of the sentence <1>. I found that the chunk to the right of the verb <regulates> of the sentence <1>. is the chunk <the mdm2 gene > and since its first token is not an entity, I tested the rest of the tokens. The entity <mdm2> is one of the tokens of the chunk <the mdm2 gene>. Which is the chunk to the right of the verb <regulates> of the sentence <1>. I found that e entity <mdm2> is one of the tokens of the chunk <the mdm2 oncogene >. Which is the chunk to the left of the verb <inhibits> of the sentence <3>. I found that the chunk to the right of the verb <inhibits> of the sentence <3>. is the chunk <p53 mediated transactivation > and the entity <p53> is one of the tokens of the chunk <p53 mediated transactivation>. Hence, it follows that <p53> is influenced by <p53>."*

The causal relations recognized above in the text fragment processed by the system form a closed loop. The above explanation may not be convenient for a user facing a crisis like in a defense situation or in a medical emergency department due to its length. A shorter one could be generated as follows:

> *"I found that the entity <p53> occurs at the left of the verb of sentence <1> and that the mdm2 gene occurs at the right of the verb of <1> and that mdm2 occurs at the left of the verb of <3> and that p53 occurs at the right of the verb of <3>; hence, it follows that <p53> is influenced by <p53>."*

11. Discussion

Although I enjoyed playing with the two 2012 and 2013 Loebner Turing test champions, I will now present my own serious reservations concerning the Turing test. My reservations must be read in view of the 2014 claim that a program "passed" the test be presented in detail below. In my view, some of the shortcomings of the Turing test as organized for the Loebner Prize contest that lessen the value of its results with respect to evaluating artificial intelligence and differentiating from human intelligence and consciousness are:

a. It is not guaranteed that the judges have the capability of posing questions appropriate for differentiating humans from state-of-the-art computer systems.

b. The subject matter of the conversations is so wide that very often a human may be unable to answer a question simply because she has no knowledge of the topic that the question refers to.

The questions appropriate for uncovering the nonhuman nature of a computer system should utilize knowledge of at least some of the following human capabilities that usually are manifested during the dialog of a human with a judge:

1. Memorization of previous stages of a dialog.

2. Context consciousness.

3. Logical reasoning.

4. Humor appreciation.

5. Irony appreciation.

6. Metaphor understanding.

7. Creativity.

These human capabilities inspire artificial intelligence researchers in their efforts to advance the relevant technology. It is expected that programs written by programmers unfamiliar with the state of the art would fail to answer correctly questions that utilize these capabilities. In my two experiments reported above, I checked mainly for operational consciousness in line with our book's [6] main theme and the results show that this was sufficient for swiftly uncovering the machine nature of the two systems I interacted with. In addition, some lack of coherence can be observed apparently as a result of lack of dialog memorization by these systems.

I performed my experiments during the 2012–2013 period. I was subsequently surprised to learn that in the 2014 Loebner contest with the Turing test a program named "Eugene Goostman" succeeded to "pass" the test. The 2014 Turing test contest was run in London under the auspices of the University of Reading and the British Royal Society.

Cybernetics professor Kevin Warwick of the University of Reading who announced the result of the Test is reported in his Christian Monitor interview of 9th of June 2014 to state the following:

> *"Some will claim that the test has already been passed. The words Turing test have been applied to similar competitions around the world. However, this event involved the most simultaneous comparison tests than ever before, was independently verified and, crucially, the conversations were unrestricted. A true Turing test does not set the questions or topics prior to the conversations. We are, therefore, proud to declare that Alan Turing's test was passed for the first time on Saturday."*

The Christian Monitor's reporter commented:

"Despite Prof. Warwick's praise, a conversation with Mr. Goostman is decidedly underwhelming. He often appears not to be listening, fails to answer direct questions, and is inappropriately sarcastic and aggressive.

So, can machines think? We posed this question to Goostman and got an uncharacteristically direct answer. 'Machines can not think" Goostman told the Monitor."

The abovementioned difference of opinion between the media and scientists became most apparent in the wave of publicity and reactions that followed the announcement. The reactions ranged from naïve statements such as "supercomputer" first to pass Turing test, convinced judges it is alive to sober scientific analyses in line with my reservations presented above that tried to rationalize the event. The writer of the above mentioned "headline" falls victim of the usual confusion between the word "computer," which is used to refer to the general-purpose stored-program digital electronic device with the phrase "computer system" that refers to a computer equipped and running executing the instructions of a "program". This set of computer instructions, when executed by the computer, makes it behave as a special-purpose computer. When on September 15th, 2014, I googled the name of the 2014 prize winner program I got 125.000 snippets as the result of my search. This shows the wide publicity of the Turing test of the 2014 Loebner Prize contest. However, I tried in vain several times to interact with Eugene Goostman, but I always found it inactive.

I will now briefly review some of the recent scientific works that deal with the so called "Turing test" that are related to the definition of "artificial intelligence" since it is usually stated that if a program passes the test, it is then considered "intelligent" and hence satisfies the definition of "artificial intelligence." I must first emphasize that Turing in his seminal 1950 paper did not exactly propose the so called "Turing test", as organized for the Loebner Prize contest as a test of intelligence. In his own words in that paper, it is stated in page 442:

> *"I believe that in about 50 years' time, it will be possible to program computers, ..., to make them play the imitation game so well that an average interrogator will not have more than 70 percent chance of making the right identification after 5 minutes of questioning. The original question "Can machines think?" I believe it to be too meaningless to deserve discussion. ... Conjectures are of great importance since they suggest useful lines of research."*

I think that the important fact is the formulation of his conjecture, which we have to judge whether it was verified in 2014, a mere 14 years later than the year specified by his prediction. It should be noted that there are some inexact phrases in Turing's conjecture that make this verification rather hard, for example, "average interrogator," "70 percent chance," and "five minutes of questioning". The following questions, among others, must be answered before the conjecture can become precise enough for scientific scrutiny:

1. Over what population of interrogators is supposed that the "average" will be taken?

2. Will the interrogators be conversant with computer science or not?

3. How is the "70 percent chance" defined? (during an interrogators day, during the 5 minutes of questioning, or during what other time span?)

4. The five-minute time span refers to the time taken to pose the questions or to the whole duration of a dialog.

5. Will the questioning involve all human knowledge or some subset of it and in that case what subset?

Other scientists have also criticized the Turing text from both philosophical and technical points of view. Among other things, the fact is criticized that Turing did not provide a definition of the verb "think."

This is used in formulating the question "do machines think?". In view of such criticisms a new test to replace the Turing test called the "Winograd schema challenge" has been established.

The "Winograd schema challenge "has been suggested as a conceptually and practically appealing alternative to the Turing test.

The Winograd schema challenge is an alternative to the Turing test that is supposed to provide a more accurate measure of artificial intelligence. Rather than base the test on the sort of freeform conversation suggested by the Turing test, the Winograd schema challenge poses a set of multiple-choice questions.

An example of a Winograd schema question is the following:

> *"The trophy would not fit in the brown suitcase because it was too big. What was too big?*
> *Answer 0: the trophy or Answer 1: the suitcase?"*

A human who answers these questions correctly typically uses his abilities in spatial reasoning, his knowledge about the typical sizes of objects, and other types of commonsense reasoning, to determine the correct answer. The 2015 Commonsense Reasoning Symposium, to be held at the AAAI Spring Symposium at Stanford from March 2325, 2015, will include a special session for presentations and discussions on progress and issues related to this Winograd schema challenge. Contest details can be found at: http://commonsensereasoning.org/winograd.html.

A program succeeding in the Winograd schema challenge needs reasoning abilities and background knowledge. It involves a coreference resolution task. The complexity of the task is increased by the fact that the Winograd sentences are not constrained by domain or sentence structure.

The winner program that meets the baseline for human performance will receive a grand prize of $25,000. In the case of multiple winners, a panel of judges will base their choice on either further testing or examination of traces of program execution. If no program meets those thresholds, a first prize of $3000 and a second prize of $2000 will be awarded to the two highest-scoring entries. In the case of teams, the prize will be given to the team lead whose responsibility will be to divide the prize among their teammates as appropriate.

In [13] a pronoun resolver system is developed for the confined domain Winograd sentences. A classifier or filter was developed which takes input sentences and decides to accept or reject them based on some criteria.

Furthermore, he has developed four answering modules, which use world knowledge and inference mechanisms to try and resolve the pronoun.

In [14] they examine resolving complex cases of definite pronouns, specifically those for which traditional linguistic constraints on coreference, as well as commonly used resolution heuristics are not useful.

In [15] a test quite different from both the Turing test and the Winograd challenge is studied, namely the "Raven's Progressive Matrices" (RPM) Test. This test is based upon purely visual representations. A technique is introduced based on the calculation of confidence in an answer and the automatic adjustment of level of resolution if that confidence is insufficient.

12. Conclusions

As shown the 2012 and 2013 winners of the Loebner Prize can easily be disclosed as non-human as they display almost zero consciousness and sense of dialog cohesion. The many analysis dialogs with GPT showed that it can respond usually correctly where the said champions failed to. However, analysis dialogs with more abstract areas, some limitations of GPTs were shown as expected since they use statistics and lack full logical reasoning. In [16–20] their limitations for intelligent question answering and similar tasks are examined.

Author details

John Kontos
National and Kapodistrian University of Athens, Athens, Greece

*Address all correspondence to: ikontos2017@gmail.com

References

[1] Turing A. Computing machinery and intelligence. Mind. 1950;**LIX**:433-460

[2] Green JR et al. Baseball: An automatic question-answerer. Western Joint IRE-AIEE-ACM Computer Conference. 1961: 219-224

[3] Kontos J. ARISTA: Knowledge engineering with scientific texts. Information and Software Technology. 1992;**34**:611-616

[4] Kontos J. Machine consciousness and question answering. American Journal of Data Mining and Knowledge Discovery. 2016;**1**:7-15

[5] Kontos J. Human-system interface with explanation of actions for autonomous anti-UAV systems. International Journal of Artificial Intelligence & Applications. 2021;**12**(4):51-58

[6] Kontos J, Kasda P. Artificial Intelligence Professor John Kontos Needles Poly Kasda's Conscious Eye. Athens, Greece: Published by Notios Anemos; 2021. pp. 1-297

[7] de Gennaro M et al. Effectiveness of an empathic chatbot in combating adverse effects of social exclusion on mood. Frontiers in Psychology. 2020;**10**:3061. DOI: 10.3389/fpsyg.2019.03061

[8] Schweizer Pl. The externalist foundations of a truly total turing test. Minds and Machines. 2012;**22**: 191-212

[9] Penco C. Updating the Turing test. Wittgenstein, turing and symbol manipulation. Open Journal of Philosophy. 2012;**2**(3):189-194

[10] Shieber SM. There Can Be No Turing Test Passing Memorizing Machines. Philosophers. 2014;**14**(16):113

[11] McDermott D. On the claim that a table lookup program could pass the turing test. Minds and Machines. 2014; **24**(2):143-188

[12] Edmonds B, Gerchenson C. Learning, social intelligence and the turing test. Invited talk at turing centenary conference special session on "The turing test and thinking machines". Published in Lecture Notes in Computer Science. 2012:182-192

[13] Budukh TU. An intelligent coreference resolver for Winograd schema sentences containing resolved semantic entities. In: Thesis Presented in Partial Fulfillment of the Requirements for the Degree Master of Science Approved. Arizona, USA: Arizona State University; 2013

[14] Rahman A, Ng V. Resolving complex cases of definite pronouns: the Winograd schema challenge. In: Proceedings of the 2012 Joint Conference on Empirical Methods in Natural Language Processing and Computational Natural Language Learning. USA: Association for Computational Linguistics; 2012

[15] McGreggor K, Goel A. Confident reasoning on Raven's progressive matrices tests. In: Proceedings of the 28th AAAI Conf. on Artificial Intelligence. 2014

[16] Chai A et al. EAGER: Explainable question answering using knowledge graphs. In: 6th Joint Workshop on Graph Data Management Experiences & Systems and Network Data Analytics, Seattle, USA. 2023

[17] Liu H et al. Evaluating the logical reasoning ability of ChatGPT and GPT-4. arXiv:2304.03439v3 [cs.CL] 5. 2023. pp. 1-11

[18] Binza M, Schulza E. Using cognitive psychology to understand GPT-3. Proceedings of the National Academy of Sciences of the United States of America. 2023;**120**(6):1-10

[19] Yang LH et al. ChatGPT is not enough: Enhancing large language models with knowledge graphs for fact-aware language modeling. arXiv: 2306.11489v1 [cs.CL]. 2023. pp. 1-20

[20] Gewirtz D. Can AI detectors save us from ChatGPT? I tried 3 online tools to find out. ZDNET. 2023. Available from: https://www.zdnet.com/

Chapter 4

Multilingual Chatbots to Collect Patient-Reported Outcomes

Matej Rojc, Umut Ariöz, Valentino Šafran and Izidor Mlakar

Abstract

With spoken language interfaces, chatbots, and enablers, the conversational intelligence became an emerging field of research in man-machine interfaces in several target domains. In this paper, we introduce the multilingual conversational chatbot platform that integrates Open Health Connect platform and mHealth application together with multimodal services in order to deliver advanced 3D embodied conversational agents. The platform enables novel human-machine interaction with the cancer survivors in six different languages. The platform also integrates patients' reported information as patients gather health data into digital clinical records. Further, the conversational agents have the potential to play a significant role in healthcare, from assistants during clinical consultations, to supporting positive behavior changes, or as assistants in living environments helping with daily tasks and activities.

Keywords: embodied conversational agents, multimodal sensing, artificial intelligence, spoken language interfaces, cancer survivors

1. Introduction

An important type of patient-gathered health data (PGHD) represents so-called patient-reported outcomes (PROs). They are in general collected from patients in order to help address a health concern [1] and represent self-reports from everyday life. Therefore, in healthcare, they are also important data sources [2]. Further, PROs have become a complementary data source to telemonitoring [3], data mining, and imaging-based AI techniques [4–8]. Nowadays, the knowledge domains of clinical specialties are expanding rapidly, while due to the sheer volume and complexity of data, clinicians often fail to really exploit its potential [9]. Firstly, patient outcomes were collected mostly face to face, using paper-written forms [10–12]. Forms were added to paper-form health records (HRs), and only after the advances of information and communication technologies (ICT), the HRs are slowly being digitalized. Several studies already showed the efficiency of electronic questionnaire apps on, e.g., smartphones [13, 14]. Thus, electronic PROs, supported by artificial intelligence techniques, can further improve dropout and acceptance-rates. Further, they are also able to improve clinical and patient "satisfaction" [15–17]. A perfect example of how patient gathered health data (PGHD) and PROs are able to improve quality of life (QoL) is, e.g., ambient assisted living (AAL). Namely, AAL environments already exploit mobile devices, smart home products, software applications, and other wearable devices in the individual's everyday environment [17, 18].

Significant advances in speech and natural language processing (NLP) technologies already offer more personalized and human-like interaction, i.e., symmetric multimodality. Therefore, several spoken language interfaces, chatbots, and enablers, and the conversational intelligence became an emerging field of research in man-machine interfaces based on artificial intelligence techniques. Thus, embodied conversational agents (ECAs) can play an important role in healthcare, e.g., assistants in AAL environment in order to help with activities and daily tasks, or assistants during clinical consultations, in order to support positive behavior changes [19, 20]. These advanced interactive systems may certainly have a major impact on long-term sustainable quality of results and patient adherence over time.

The main challenges represent interoperability, integration of PGHD data, and lack of standardization [21, 22]. Namely, in healthcare, the integration of PGHD data in clinical decision-making still presents a big problem. Further, in the interoperability of electronic health records (EHRs), the unified representation of electronic health records (EHRs) still represents an issue. In order to get the highest contribution from PROs and PGHD, we considered the following: (i) "how to integrate data into clinical workflow?", (ii) "the cost and time for collecting PROs?", (iii) "how to efficiently collect data from patients?", and (iv) "how to enable proper interpretation by the clinicians?"

Within a Horizon 2020 project (PERSIST, https://projectpersist.com/, last accessed 19 June 2021), therefore, we propose a holistic system for collecting PROs remotely via both multilingual chatbots and ECAs. Further, the integration of PROs into the clinical workflow by using FHIR has been proposed. The FHIR server is located at the Open Health Connect (OHC) platform, and all traffic is orchestrated by a so-called multimodal sensing network (MSN) that runs several microservices, such as PLATTOS text-to-speech (TTS) system, ECA, RASA-based chatbot system, and SPREAD automatic speech recognition (ASR) system. In this way, we offer a fully symmetric model of interaction supporting speech, gesture, and facial expression on input and output. Further, the FHIR methodology is delivered as an enabler for efficient integration and a fully functional FHIR server [23].

The paper is structured as follows: in Section 2, related works and the ideas of our study will be presented. The PERSIST platform is described in Section 3, and fully symmetric ECA-based interaction model in Section 4. The results are presented in Section 5. In Section 6, the contributions of the PERSIST system are discussed, and the paper ends with the conclusions.

2. Related works

The paradigm of value-based healthcare represents a shift toward more efficient and more effective medical care. However, it requires additional sources of data to improve shared decision-making and enable more personalized decision-making. Therefore, conversational intelligence can significantly contribute to patient activation and engagement [24]. The technology is based on spoken language technologies (SLT), i.e., NLP, ASR, chatbot, and TTS, that enables machines to interact with humans in very natural way, using mobile or web platforms [25]. In healthcare, this started already in 1966 with ELIZA [26]. Nowadays, conversational agents have been used to solve much more complex tasks, such as booking tickets and acting as customer service agents [27]. In healthcare, conversational agents can provide patients with, e. g. personalized health and therapy information and relevant products and services. Additionally, they can connect them with healthcare providers, suggest diagnoses, and even recommended treatments based on patient symptoms and reports. Namely, multilingual communication, cost-effectiveness, and 24/7 availability make embodied conversational agents (ECAs) very useful for all those patients who have major medical concerns outside of doctor's operating hours. Several studies show that patients can perceive ECAs as interaction partners instead of human physicians and are able to trust them. Thus, they are willing to disclose medical information report more symptoms, etc. [28]. In oncology setting, CI (conversational intelligence) focuses mostly on (speech-enabled) chatbots [29]. They can contribute to lifestyle changes [30], to screening (i.e., iDecide [31]) and improving mental health state through managing psychological distress [32–34]. Therefore, chatbots are already well recognized as an enabler for adherence, active patient engagement, and satisfaction increase [35, 36]. However, the chatbots still tackle the long-term adherence with sustainable quality of the reported data [37]. In [36], they reported that active use of this technology drops already after 14 days. Namely, patients' understanding, their ability to remember the details, and perceived trustworthiness are the main factors of patient adherence [38]. Therefore, in the system of the PERSIST project, an ECA is additionally introduced. ECAs can undoubtedly increase this long-term adherence by engaging with users in interaction that is enriched by incorporating nonverbal communication [37]. Since ECA is autonomous and intelligent software entity with an embodiment used to communicate with the user [39], it can provide a system with symmetric multimodality based on speech, gesture, and facial expression. Embodiments can be designed as virtual human characters, animals, or robots [40–42]. Such fully symmetric interaction opens up the opportunity to introduce human-like qualities and significantly improves the believability of the human-machine interfaces [43]. ECAs in healthcare can be used for the treatment of mood disorders, anxiety, psychotic disorders, autism, substance use disorders, etc. [44]. In [17], ECAs already proved a promising tool for persuasive communication in healthcare. While in [42], technological and clinical possibilities of less complex ECAs were investigated, and ECAs are also shown to be a solution for routine applications in the means of rapid development, testing, and application. Stal in [45] also found out that the agents' textual output and/or speech as well as its gaze and facial expressions are the most important features. In general, for healthcare, ECA studies focused mainly on physical activity [46–48], stress [30], nutrition [49, 50], blood glucose monitoring [41], and sun protection [51]. However, there are several other studies that focus on speech, facial, and gaze expressions as the main design features [45]. ECAs in healthcare are mostly 2D-based, since gestures and appearance are not considered as main design features, and only a few studies addressed gestures.

In the PERSIST system, therefore, we use two 3D embodied conversational agents, female or male that can interact with patients in the following six languages: Slovenian, English, Spanish, French, Russian, and Latvian. ECAs are able to represent facial expressions and exploit gestures in order to enhance user experience. Namely, in this way, it is possible to better support verbal counterparts, regulate communicative relationships, and maintain clarity in the discourse.

In [52], the conversational agents are designed as a prototype, while the contribution to health-related outcomes is evaluated without relevant statistical significance. Further, Sayeed et al. in [53] describe an approach to create a patient-centered health system that is based on the FHIR standard and applications that can make requests and reports of HL7 FHIR resources.

3. The multilingual ECA-based PERSIST platform

3.1 The multilingual sensing network (MSN)

In **Figure 1**, we present the building blocks of the MSN network. The MSN consists of Apache Camel module. This module implements ActiveMQ Artemis, REST API, and Apache Kafka. In this way, we implemented specific machine-to-machine (M2M) communication between several services. The ActiveMQ Artemis module is then used for the MQTT broker. And Apache Kafka module is used for microservice architecture. The Apache Camel module is like a router in the system, since it has the ability to convert asynchronous to synchronous messages, or vice versa. We can run Apache Camel module also as a Spring Boot application in order to provide REST API end points for all HTTP requests. The MQTT broker in the system represents a link between mHealth app and OHC. Namely, the mHealth app is MQTT client that is just subscribed to ActiveMQ Artemis module. Further, microservices are using HTTP APIs and Kafka topics. For microservices, asynchronous communication is used. All predefined topics for dedicated language are supported. The synchronous communication is then used for RASA chatbot. In this case, HTTP REST requests are used and performed via Camel REST end points API.

In **Figure 2**, we can recognize two types of connections. The first one represents the synchronous connection used for communication over the secured application protocol HTTPS REST. It is needed for questionnaires, responses, and requests. The second one is then asynchronous connection. It is needed for the MQTT protocol, where we use MQTT topics. Established connections with the OHC platform can use

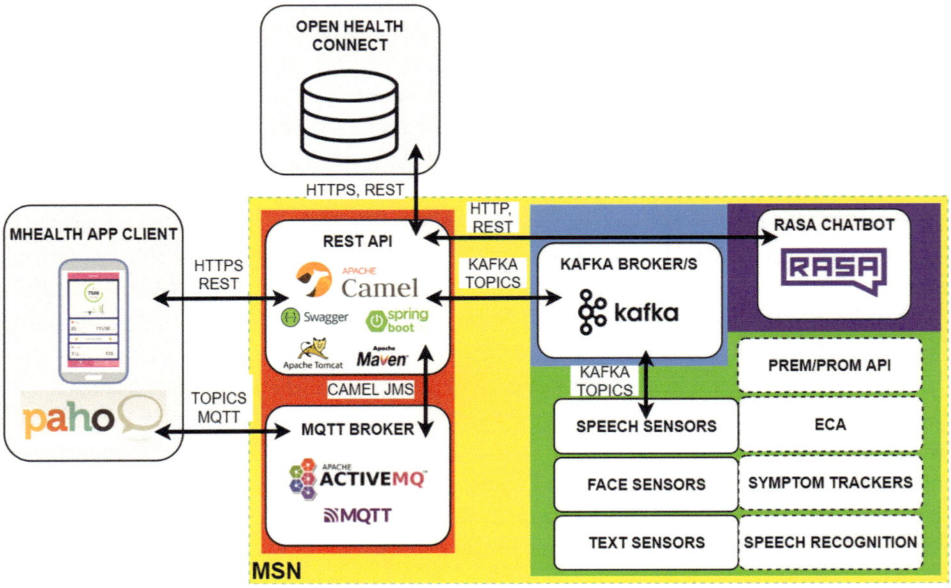

Figure 1.
The architecture of the PERSIST system.

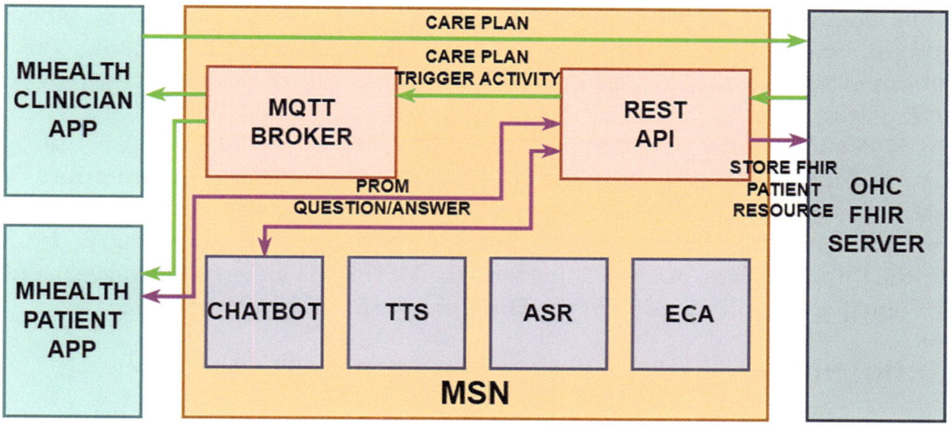

Figure 2.
Machine-to-machine communication (M2M) platform for the PERSIST system.

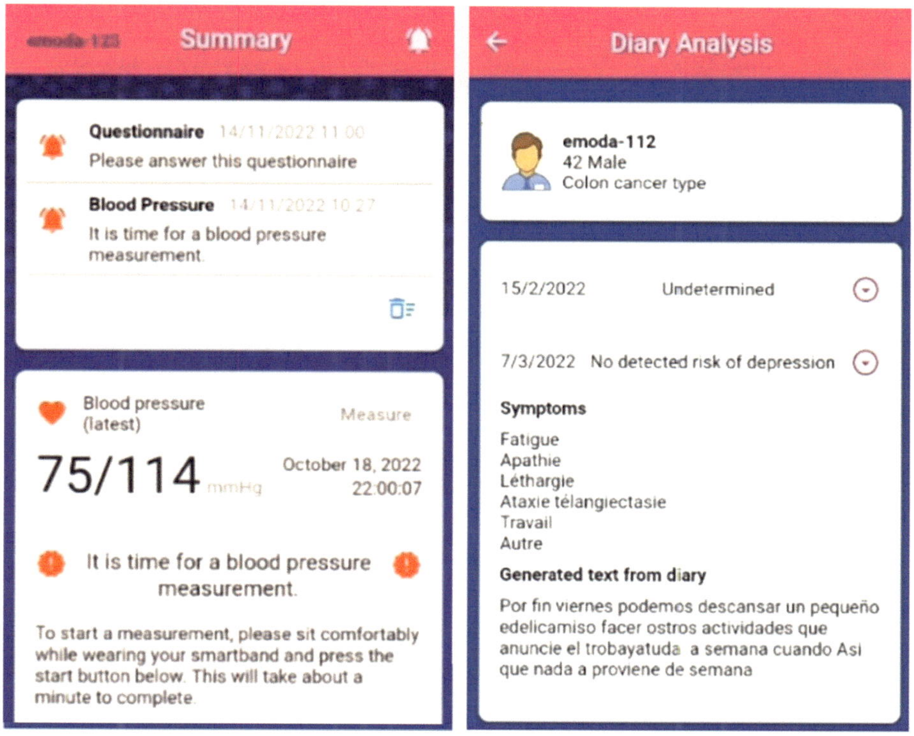

Figure 3.
Patient (left) and clinician (right) mHealth app interface.

in this way synchronous HTTPS REST protocol. Further, MSN internal connections uses MQTT, Camel Java Messaging Service (or JMS), Kafka topics, and REST.

3.2 mHealth application

As can be seen in **Figure 3**, patients and clinicians have separate mHealth applications. One is patient mHealth application that is used for data gathering and trends monitoring, while the clinician mHealth application is used for patient monitoring and specifying the patient's care plans (developed by company Emoda). The first one enables mood selection, diary recordings, reading of specific articles advised by clinicians, etc. And the second one has options to see the patients' lists and their clinical details. It is also possible to delete or edit existing patient records, or create a new one. Further, new appointments can be created by clinicians, receive notifications from patients, see the calendar, or just send/receive messages from patients. Thus, this application uses both asynchronous and synchronous protocols. We use the REST protocol for communication with the MSN REST OpenAPI (Swagger) and OHC end points, and for receiving notifications the MQTT protocol is used.

3.3 OHC FHIR server

The OHC platform has been provided by Dedalus. Basically, this is a streaming and integration platform that can be used for large--scale distributed environments. This digital health platform can also unlock isolated data. Further, OHC enables all the interfaces to be connected to and make decisions across disparate data sources in real time. It comprises a set of components, as depicted in the conceptual/logical architectures, is flexible, and can be deployed on private data center, or via cloud in environments like Azure or AWS. It provides the latest version of HAPI FHIR R4 [54].

4. The fully symmetric ECA-based interaction model

4.1 End-to-end multilingual text-to-speech synthesis system PLATTOS

Text-to-speech (TTS) PLATTOS in **Figure 4** is the first microservice in the PERSIST system. It is used for generating speech from text for the ECA agents that communicate with the patients. The PLATTOS system follows ideas presented in [55, 56] and enables real-time generation of speech in several languages, with practically human-like quality. It is basically the combination of two complex network models: a feature prediction NN model and a flow-based neural-network-vocoder WaveGlow.

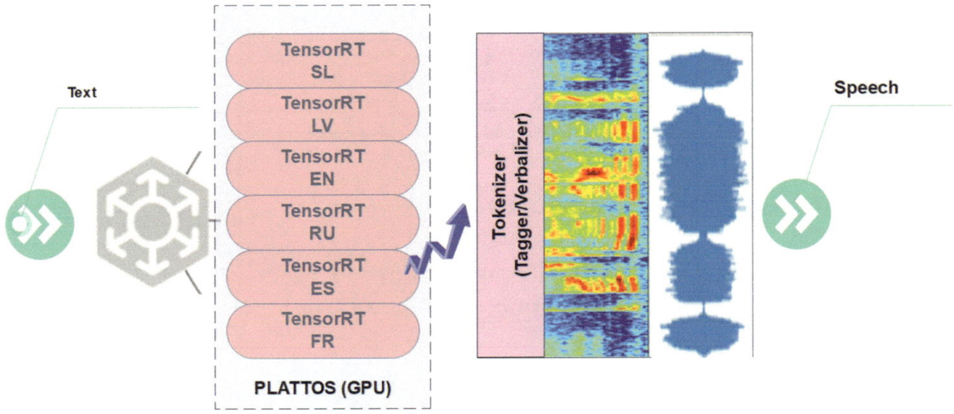

Figure 4.
TTS system PLATTOS.

4.2 End-to-end multilingual speech recognition system SPREAD

This microservice is developed to support the spoken language-based interface in the Health app and to feed the survivor's answers to the dialog management component (i.e., RASA chatbot) for several languages. E2E ASR system SPREAD in **Figure 5** follows some ideas from Jasper model [57–59], where the training has been improved by NovoGrad optimizer.

4.3 Embodied conversational system and embodied conversational agent

A RASA NLU [60] and ECA framework [61] are a core framework for an Embodied Conversational System (ECA). In this way, multilingual ECAs are capable of creating responses in natural language. All responses can also be visualized. Namely, multilingual chatbots are used to manage the more natural discourse between the system and patient. They are implemented as an API. Here, the NLU is the main engine of the chatbots and is programmed in Python and YAML language. Chatbots are all running on a Linux server. It implements standardized patient-reported outcomes (PROs) as storylines in six languages used in the PERSIST Clinical Study [62]. For storing the data, SQLite database within RASA is used, while POST and GET requests are used to store information, such as patients' answers, questionnaires, and other events that are triggered in a specific conversation.

The ECA framework is then used to transform plain text generated by the chatbot into ECA's multimodal responses incorporating gestures. The proprietary algorithm proposed in [61] has been used (**Figure 6**). It uses proprietary EVA-Script notations.

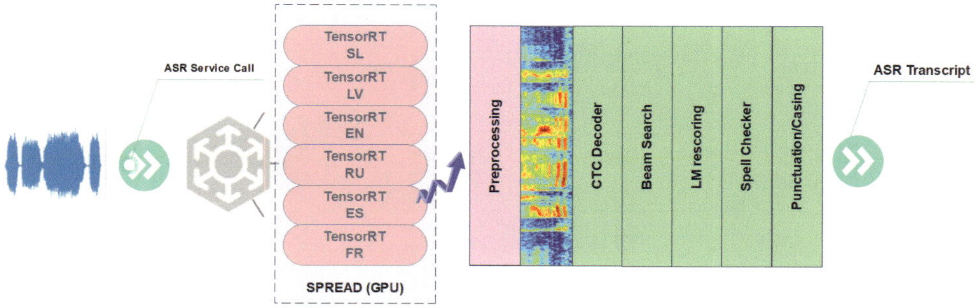

Figure 5.
ASR system SPREAD.

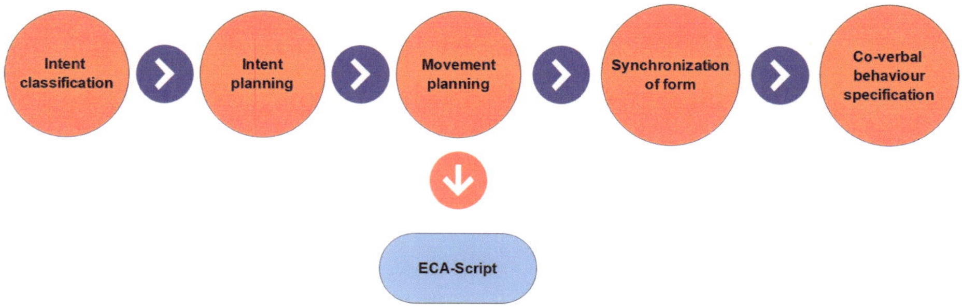

Figure 6.
Generation of expressive co-verbal behavior.

Each movement is formalized as a simultaneous execution within the block <bgesture>. The poses are described then within stroke phases, where the preparation phases are defined by <unit> blocks. Each <unit> also contains the complete configuration of individual movement controllers that are used in the representation of the specific pose. The retraction and hold phases then represent the shape being withheld or just retracted into some neutral state. They are both added within the <unit> by using attributes DurationHold and DurationRetraction.

5. Results

The PERSIST platform was deployed on two physical servers at the University of Maribor, FERI. The functional scheme of the system is highlighted in **Figure 7**. The PERSIST system is used mainly by the clinician. Namely, they have to define and schedule activities as part of patient's care workflow (phase 1). On the other hand, the patients execute activities (phase 3). MSN and OHC are the main services within the system. The MSN service is used to implement activities and make their execution more natural by delivering the symmetric model of interaction, and the OHC service is used to store data and automate the execution of the clinical workflow.

Questionnaires are available in six different languages: Slovenian, English, Russian, Latvian, French, and Spanish. On the output side, the system represents the information generated by chatbot as female ECA Eva and the male ECA Adam (**Figure 8**). In this way, in the output also non-verbal elements are associated with synthesized speech. In this way, raw texts are presented to the user as a multimodal output, which combines a spoken communication channel and synchronized visual communication

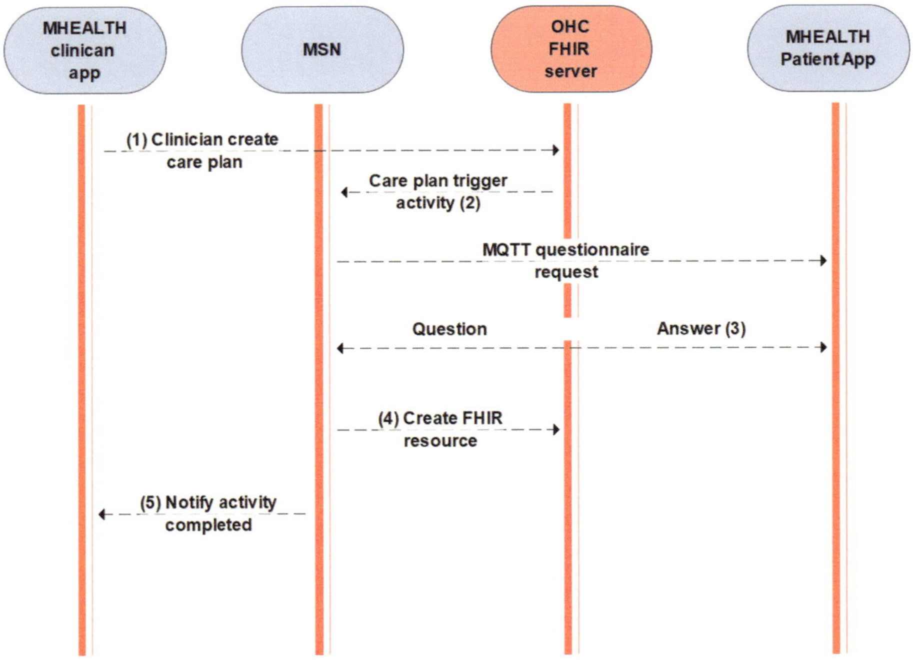

Figure 7.
Functional flow: Integration phases—Allocation of an activity (1), request for execution of the activity (2), implementation of the activity (3), creation of resource (4), and completion of the activity (5).

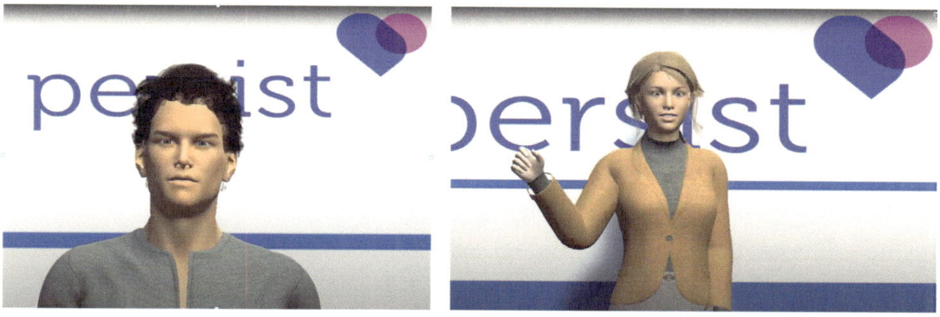

Figure 8.
Multimodal conversational response with ECAs.

channel. At the input, the system accepts speech or text. Additionally, a word-to-concept mapping is delivered as part of spoken language understanding. This is needed in order to properly map user responses into answers expected by PROs.

We deployed the system on a server hosting five virtual machines over the Proxmox VE 6.3–2. Further, the server is running the Xubuntu 20.04 LTS operating system. On the other platform, named PERSIST_INFERENCE, there are the Ubuntu Server 20.04 LTS OS, and microservices for ASR, TTS, and ECA. Microservices are integrated using predefined topics, and Kafka producers and consumers. To evaluate the hardware performance of the system, we simulated the load on the system by measuring CPU usage, memory usage, and average response time for both Camel and RASA chatbot. The results are outlined in **Figures 9-11**.

As seen in **Figure 9**, with the duplication of active users in tests the CPU usage is rising linearly from 11.65% with 25 active users to 56.04% with 1000 active users in the case of Camel, and mostly linear from 5.86% with 25 active users to 30.44% with 1000 active users for Rasa chatbot. The volatile memory was stagnating on both

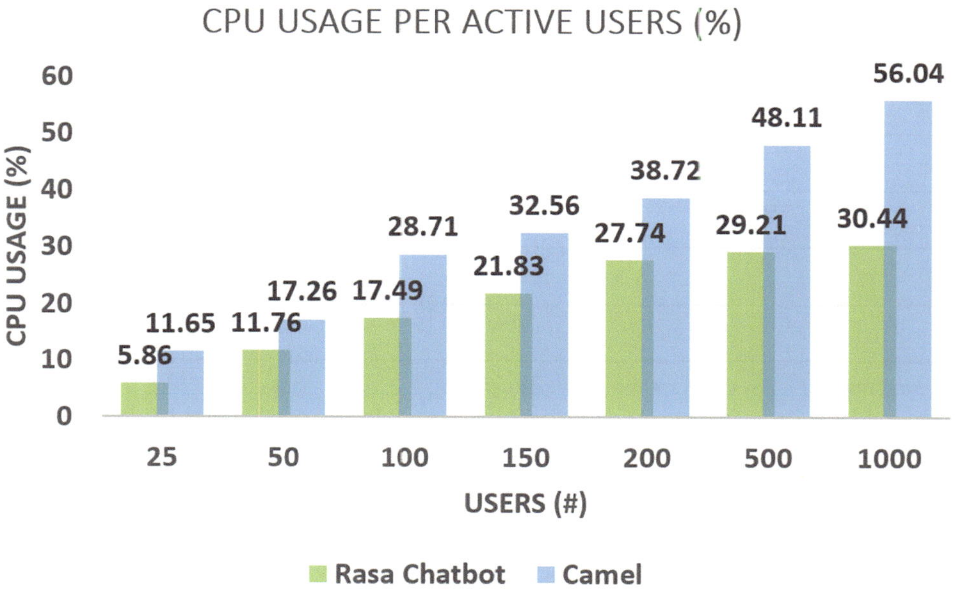

Figure 9.
CPU use (%) per active users.

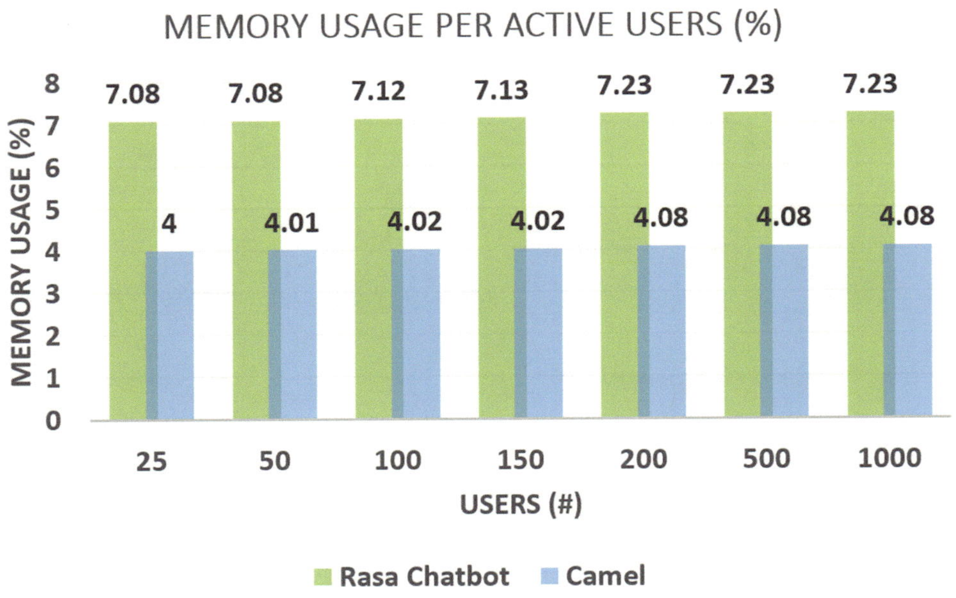

Figure 10.
Memory consumption (GB) per active users.

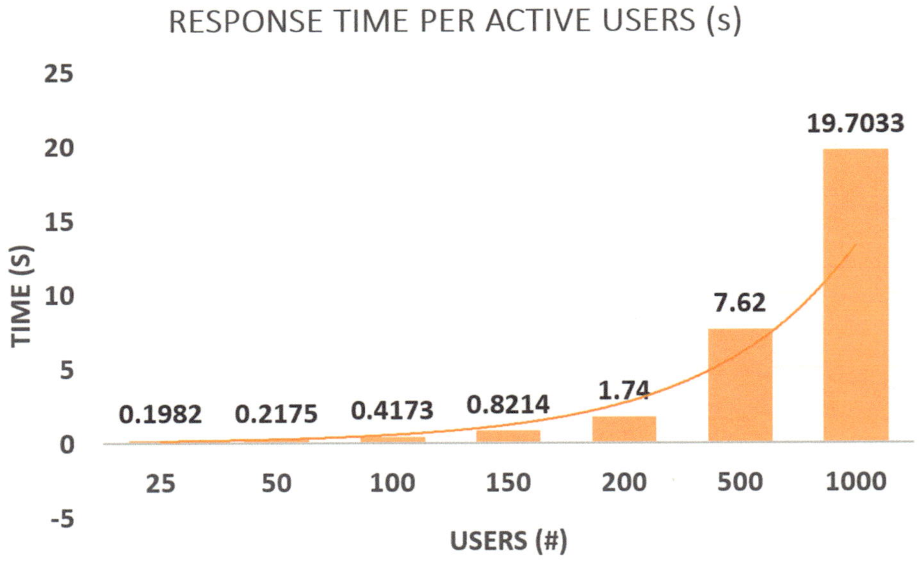

Figure 11.
Graphical results of average response time per active user.

the Camel and the Rasa chatbot and proved independent of the increase of users (**Figure 10**). In the case of the Camel, the memory usage was near 50%, while on the Rasa chatbot near 25%. Further, **Figure 11** presents the MSN's internal average response time on requests between 25 and 1000 active users. The response time in this case is 0.1982 s with 25 active users and is increasing linearly as the number of users is increasing. We have 1.74 s response time with 200 active users. Then it starts rising more exponentially to 197,033 s delay, with 1000 active users.

The models for the end-to-end ASR system SPREAD for six languages were trained on DGX-1, 8 × V100, 8 × 32 g GPU MEM, while the inference engine had 2 RTX8000, with 2 × 48 g GPUMEM. The audio datasets size used was minimal 1700 h of speech. The best model reached 2.6% WER, and all other models reached below 9% WER. The quality of the end-to-end TTS system PLATTOSand MUSHRA listening tests [63] were performed by PERSIST consortium partners. In this way, 21 consortium members participated, all in general with background knowledge in this field. Different TTS architectures were evaluated, while the architecture based on Tacotron and Waveglow was best rated. PLATTOS for all six languages was evaluated with score around 82 on 100 level scale. The results show that speech generated is highly intelligible and understandable. Further, the evaluation of the multimodal conversational response was reported in [61], where 30 individuals assigned an average score of 3.45 on the five-level Likert scale. The results show that the system produces a very viable and believable natural user interface.

6. Discussion

The main challenges for wide adaptation of PGHD in clinical practice include usability and sustainable quality of results (i.e., patient motivation and adherence) [21, 37]. The presented system includes patient/clinician mobile applications, OHC FHIR server, and the MSN server. OHC FHIR server provides interoperability between all components. The framework provides several tools that can be used for ingestion, indexing, storage, integration, and surfacing of patient information. In this way, the PERSIST system represents an open digital integration hub that can deliver scale, speed, and flexibility to securely gain value through the integration of health systems. Further, the OHC enables innovation through near-real-time access to longitudinal patient records, where the APIs provide opportunities to flexibly design services that can seamlessly ingest discrete data from the source into a third-party application. The FHIR has also been recognized as an approach suitable for citizen developers, since it also supports "low-code/no-code" solutions [21]. Our future efforts will be directed toward transformation and ingestion of EHRs from existing IT platforms into FHIR ready server. Based on the studies, the main activities will involve the definition of an ontology that will correlate existing fields with specific FHIR resources. The information in existing EHRs is mostly stored as partially structured or unstructured text; therefore, a specific focus will be directed toward extracting information by using modern NLP techniques and data to concept mapping.

The other challenge relates to the patient's perspective and long-term sustainability and quality of collected information [36, 37]. Perceived complexity and trustworthiness represent also the main drivers of patient adherence [38]. Therefore, MSN delivers the necessary microservice infrastructure, where the services are distributed among the servers and can be replicated if needed. A fully articulated ECA was deployed for all six languages in order to implement more natural human-machine interaction, where the EVA realization framework transforms the co-verbal descriptions contained in EVA events into articulated movement generated by the expressive virtual entity. The EVA-Script language is actually applied onto the articulated 3D model EVA in the form of animated movement [43]. Trustworthiness is a clinical value, which has a significant impact on adherence mitigating pervasive threats to health [64]. The symmetric multimodal model for dialog systems enables the ECAs to deliver and to understand input/output modes, including speech, gestures, and facial expressions. This makes the interfaces more familiar and trustworthy [38], where trustworthiness is one of the building blocks of patient compliance and responsiveness [65].

The RASA chatbot API is using PREMs and PROMs to see the patients' health status and the patients' perceptions of their experience while receiving treatment. In this case, we created several stories that contained probable conversations with patients. These are basically the intents that have to be executed, based on patient's responses [66]. Inclusion of multilingual ECAs have positive effect on patient adherence, as also several experiments imply. Further, ECAs contribute to long-term sustainability and familiarity [29] and decrease the complexity of user interfaces. Namely, having a virtual body that shows the nonverbal cues can provide easier understanding of the context, coherence for information exchange, and an increase for believability and trustworthiness to the virtual entity.

However, the phenomenon of "uncanny valley" may have significant negative impact on the overall user experience with articulated entities compared to "disembodied" agents as suggested in [67]. Thus, in the future, we will focus specifically on the synchronization issues of nonverbal behavior with speech.

7. Conclusions

In this paper, a multilingual holistic approach toward sustainable collection of PGHD and PROs and their efficient integration into clinical workflow has been presented. Namely, the PGHD may contribute to personalized care and early identification related to psychological and physiological symptoms and negative health outcomes. The PERSIST system represents an opportunity to integrate the benefits and deliver them to the patients. The system consists of patient/clinician mobile applications, an OHC FHIR server, and a MSN server. The research and this study address several technologies from the prototype (proof-of-concept) perspective. The used technology was evaluated on modular basis, statistically, and on a short-term-use basis.

Acknowledgements

This study is part of the project "PERSIST: Patient-centered survivorship care plan after cancer treatment" that has received funding from the European Union's Horizon 2020 research and innovation program (GA No. 875406). This work is partially financed by the HosmartAI project, funded by European Union's Horizon 2020 research and innovation programme under grant agreement No 101016834.

Conflict of interest

The authors declare no conflict of interest. The funders had no role in the design of the study, in the collection, analyses, or interpretation of data, in the writing of the manuscript, or in the decision to publish the results.

Author details

Matej Rojc*, Umut Ariöz, Valentino Šafran and Izidor Mlakar
Faculty of Electrical Engineering and Computer Science, University of Maribor, Slovenia

*Address all correspondence to: matej.rojc@um.si

References

[1] National Health Council. What are Clinician-Reported Outcomes (ClinROs)? National Health Council. 2019. Available online: https://nationalhealthcouncil.org/coa-series-what-are-clinician-reported-outcomes-clinros/ [Accessed: June 19, 2021]

[2] Health IT, Office of the National Coordinator for Health Information Technology (ONC), US Department of Health Human Services. What are patient-generated health data?. Available online: http://healthit.gov/topic/otherhot-topics/what-are-patientgenerated-health-data [Accessed: October 15, 2019]

[3] Fauzana N, Gulcharan BI, Azhar MA, Daud H, Mohd NN, Taib I. Integrating emerging network technologies to heart rate monitoring system to investigate transmission stability and accuracy: Preliminary results. International Journal of Electrical Engineering Computer Science (EEACS). 2021;**3**:21-22

[4] Sawssen B, Okba T, Noureeddine L. A mammographic images classification technique via the Gaussian radial basis kernel ELM and KPCA. International Journal of Applied Mathematical Computer Science and System Engineering. 2020;**2**:92-98

[5] Zheng Q, Yang L, Zeng B, Li J, Guo K, Liang Y, et al. Artificial intelligence performance in detecting tumor metastasis from medical radiology imaging: A systematic review and meta-analysis. EClinical Medicine. 2021;**31**:100669

[6] Inès A, Zgaya H, Slim H. Workflow tool to model and simulate patients paths in Pediatric Emergency Department. International Journal of Electrical Engineering Computer Science. 2020;**2**:73-78

[7] Abdelnabi MLR, Jasim MW, El-Bakry HM, Taha MHN, Khalifa NEM, Loey M. Breast and colon cancer classification from gene expression profiles using data mining techniques. Symmetry. 2020;**12**:408

[8] Austin E, LeRouge C, Hartzler AL, Segal C, Lavallee DC. Capturing the patient voice: Implementing patient-reported outcomes across the health system. Quality of Life Research. 2020;**29**:347-355

[9] Groccia MC, Guido R, Conforti D. Multi-classifier approaches for supporting clinical decision making. Symmetry. 2020;**12**:699

[10] Ellwood PM. Outcomes Management. The New England Journal of Medicine. 1988;**318**:1549-1556

[11] Tarlov AR, Ware JE, Greenfield S, Nelson EC, Perrin E, Zubkoff M. The medical outcomes study: An application of methods for monitoring the results of medical care. JAMA. 1989;**262**:925-930

[12] Bielli E, Carminati F, La Capra S, Lina M, Brunelli C, Tamburini M. A wireless Health outcomes monitoring system (WHOMS): Development and field testing with cancer patients using mobile phones. BMC Medical Informatics and Decision Making. 2004;**4**:1-13

[13] Tran C, Dicker A, Leiby B, Gressen E, Williams N, Jim H. Utilizing digital health to collect electronic pa-tient-reported outcomes in prostate cancer: Single-arm pilot trial. Journal of Medical Internet Research. 2020;**22**:e12689

[14] Wright AA, Raman N, Staples P, Schonholz S, Cronin A, Carlson K, et al. The HOPE pilot study: Harnessing patient-reported outcomes and biometric data to enhance cancer care. JCO Clinical Cancer Information. 2018;**2**:1-12

[15] Rajguru P, Ryan S, McLaurin E, Wirta D, Grieco J. A novel method for collecting patient reported outcomes (PROs): Developing and validating electronic PROs on a mobile smartphone platform. Investigative Ophthalmology & Visual Science. 2020;**7**:110

[16] Van Egdom LSE, Pusic A, Verhoef C, Hazelzet JA, Koppert LB. Machine learning with PROs in breast cancer surgery; caution: Collecting PROs at baseline is crucial. The Breast Journal. 2020;**26**:1213-1215

[17] Kramer LL, Ter Stal S, Mulder B, De Vet E, Van Velsen L. Developing embodied conversational agents for coaching people in a healthy lifestyle: Scoping review. Journal of Medical Internet Research. 2020;**22**:e14058

[18] Queirós A, Dias A, Silva AG, Rocha NP. Ambient assisted living and health-related out-comes-a systematic literature review. Inform. 2014;**4**:19

[19] Alosaimi W, Ansari TJ, Alharbi A, Alyami H, Seh A, Pandey A, et al. Evaluating the impact of different symmetrical models of ambient assisted living systems. Symmetry. 2021;**13**:450

[20] Laranjo L, Dunn A, Tong HL, Kocaballi AB, Chen J, Bashir R, et al. Conversational agents in healthcare: A systematic review. Journal of the American Medical Informatics Association. 2018;**25**:1248-1258

[21] Jim HSL, Hoogland A, Brownstein NC, Barata A, Dicker AP, Knoop H, et al. Innovations in research and clinical care using patient-generated health data. CA: A Cancer Journal for Clinicians. 2020;**70**:182-199

[22] Rehman A, Naz S, Razzak I. Leveraging big data analytics in healthcare enhancement: Trends, challenges and opportunities. Multimedia Systems. 2021;**2021**:1-33

[23] Resourcelist—FHIR v4.0.1. Available online: http://hl7.org/fhir/resourcelist.html [Accessed: April 1, 2021]

[24] Wald JS, Sands DZ. Transforming Health care delivery through consumer engagement, Health data transparency, and patient-generated Health information. Yearbook of Medical Informatics. 2014;**23**:170-176

[25] Shawar B, Atwell E. Chatbots: Are they really useful? LDV Forum. 2007;**22**:29-49

[26] Weizenbaum J. ELIZA—A computer program for the study of natural language communication between man and machine. Communications of the ACM. 1966;**9**:36-45

[27] Sharma RK, Center NI. An analytical study and review of open source Chatbot framework, Rasa. International Journal of Engineering Research. 2020;**9**:060723

[28] Palanica A, Flaschner P, Thommandram A, Li M, Fossat Y. Physicians' perceptions of Chatbots in Health care: Cross-sectional web-based survey. Journal of Medical Internet Research. 2019;**21**:e12887

[29] Bibault J-E, Chaix B, Nectoux P, Pienkowski A, Guillemasé A, Brouard B. Healthcare ex Machina: Are conversational agents ready for prime time in oncology? Clinical Translational and Radiational Oncology. 2019;**16**:55-59

[30] Gardiner PM, McCue KD, Negash LM, Cheng T, White LF, Yinusa-Nyahkoon L, et al. Engaging women with an embodied conversational agent to deliver mindfulness and lifestyle recommendations: A feasibility randomized control trial. Patient Education and Counseling. 2017;**100**:1720-1729

[31] Owens OL, Felder T, Tavakoli AS, Revels AA, Friedman DB, Hughes-Halbert C, et al. Evaluation of a computer-based decision aid for promoting informed prostate Cancer screening decisions among African American men: iDecide. American Journal of Health Promotion. 2019;**33**:267-278

[32] Fitzpatrick KK, Darcy A, Vierhile M. Delivering cognitive behavior therapy to young adults with symptoms of depression and anxiety using a fully automated conversational agent (Woebot): A randomized controlled trial. JMIR Mental Health. 2017;**4**:e19

[33] Inkster B, Sarda S, Subramanian V. An empathy-driven, conversational artificial intelligence agent (Wysa) for digital mental well-being: Real-world data evaluation mixed-methods study. JMIR mHealth and uHealth. 2018;**6**:e12106

[34] Ly KH, Ly A-M, Andersson G. A fully automated conversational agent for promoting mental well-being: A pilot RCT using mixed methods. Internet Interventions. 2017;**10**:39-46

[35] Girgi A, Durcinoska I, Levesque JV, Gerges M, Sandell T, Arnold A, et al. The PROMPT-care program group eHealth system for collecting and utilizing patient reported outcome measures for personalized treatment and care (PROMPTCare) among cancer patients: Mixed methods approach to evaluate feasibility and acceptability. Journal of Medical Internet Research. 2017;**19**:e330

[36] Kneuertz PJ, Jagadesh N, Perkins A, Fitzgerald M, Moffatt-Bruce SD, Merritt RE, et al. Improvingpatientengagement,adherence, and satisfaction in lung cancer surgery with implementation of a mobile device platform for patient reported outcomes. Journal of Thoracic Disease. 2020;**12**:6883-6891

[37] Tellols D, Lopez-Sanchez M, Rodríguez I, Almajano P, Puig A. Enhancing sentient embodied conversational agents with machine learning. Pattern Recognition Letters. 2020;**129**:317-323

[38] Martin LR, Williams SL, Haskard KB, DiMatteo MR. The challenge of patient adherence. Therapeutics and Clinical Risk Management. 2005;**1**:189-199

[39] Isbister K, Doyle P. The blind men and the elephant revisited evaluating interdisciplinary ECA research. In: Ruttkay Z, Pelachaud C, editors. From Brows to Trust Evaluating Embodied Conversational Agents. Dordrecht, The Netherlands: Springer; 2004. pp. 3-26

[40] Bickmore T, Gruber A, Picard R. Establishing the computer–patient working alliance in automated health behavior change interventions. Patient Education and Counseling. 2005;**59**:21-30

[41] Klaassen R, Bul KCM, Akker ROD, Van Der Burg GJ, Kato PM, Di Bitonto P. Design and evaluation of a pervasive coaching and gamification platform for young diabetes patients. Sensors. 2018;**18**:402

[42] Provoost S, Lau HM, Ruwaard J, Riper H. Embodied conversational agents in clinical psychology: A scoping review.

Journal of Medical Internet Research. 2017;**19**:e151

[43] Rojc M, Kăčič Z, Mlakar I. Advanced content and Interface personalization through conversational behavior and affective embodied conversational agents. In: Fernandez MAA, editor. Artificial Intelligence Emerging Trends and Applications. London, UK: IntechOpen; 2018

[44] Brinkman WP. Virtual health agents for behavior change: Research perspectives and directions. In: IVA2016 Workshop: Graphical and Robotic Embodied Agents for Therapeutic Systems - GREATS16. Los Angeles, CA, USA: Institute for Creative Technologies, USC; 2016

[45] Stal S, Kramer LL, Tabak M, Akker HOD, Hermens H. Design features of embodied conversational agents in eHealth: A literature review. International Journal of Human Computer Studies. 2020;**138**:102409

[46] Friederichs S, Bolman C, Oenema A, Guyaux J, Lechner L. Motivational interviewing in a web-based physical activity intervention with an Avatar: Randomized controlled trial. Journal of Medical Internet Research. 2014;**16**:e48

[47] Bickmore TW, Caruso L, Clough-Gorr K, Heeren T. 'It's just like you talk to a friend' relational agents for older adults. Interacting with Computers. 2005;**17**:711-735

[48] Ellis T, Latham NK, DeAngelis TR, Thomas CA, Saint-Hilaire M, Bickmore TW. Feasibility of a virtual exercise coach to promote walking in community-dwelling persons with Parkinson disease. American Journal of Physical Medicine & Rehabilitation. 2013;**92**:472-485

[49] Henkemans BOA, van der Boog PJ, Lindenberg J, van der Mast CA, Neerincx MA, Zwetsloot-Schonk BJ. An online lifestyle diary with a persuasive computer assistant providing feedback on self-management. Technology and Health Care. 2009;**17**:253-267

[50] Bickmore TW, Schulman D, Sidner C. Automated interventions for multiple health behaviors using conversational agents. Patient Education and Counseling. 2013;**92**:142-148

[51] Sillice AM, Morokoff PJ, Ferszt G, Bickmore T, Bock BC, Lantini R, et al. Using relational agents to promote exercise and sun protection: Assessment of participants' experiences with two interventions. Journal of Medical Internet Research. 2018;**20**:e48

[52] Benze G, Nauck F, Alt-Epping B, Gianni G, Bauknecht T, Ettl J, et al. PROutine: A feasibility study assessing surveillance of electronic patient reported outcomes and adherence via smartphone app in advanced cancer. Annals of Palliative Medicine. 2019;**8**:104-111

[53] Sayeed R, Gottlieb D, Mandl KD. SMART markers: Collecting patient-generated health data as a standardized property of health information technology. NPJ Digital Medicine. 2020;**3**:1-8

[54] Versions—FHIR v4.0.1. Available online: https://www.hl7.org/fhir/versions.html [Accessed: April 1, 2021]

[55] Shen J, Pang R, Weiss RJ, Schuster M, Jaitly N, Yang Z, et al. Natural tts synthesis by condi-tioning wavenet on mel spectrogram predictions. In: Proceedings of the 2018 IEEE

International Conference on Acoustics, Speech and Signal Processing. (ICASSP), 15-20 April 2018. Calgary, AB, Canada: IEEE Press; 2018. pp. 4779-4783

[56] Oord AVD, Dieleman S, Zen H, Simonyan K, Vinyals O, Graves A, et al. Wavenet: A generative model for raw audio. arXiv 2016, arXiv:1609.03499.57

[57] Li J, Lavrukhin V, Ginsburg B, Leary R, Kuchaiev O, Cohen JM, et al. Jasper: An End-to-End Convolutional Neural Acoustic Model. 2019:71-75. DOI: 10.21437/Interspeech.2019-1819

[58] Graves A, Jaitly N. Towards end-to-end speech recognition with recurrent neural networks. In: Proceedings of the International Conference on Machine Learning (ICML 2014), 21-26 June 2014. Vol. 32. Beijing, China: JMLP, W&CP; 2014. pp. 1764-1772

[59] Chorowski J, Bahdanau D, Cho K, Bengio Y. End-to-end continuous speech recognition using attention-based recurrent nn: First results. In NIPS 2014 Workshop on Deep Learning. Dec 2014

[60] Bocklisch T, Faulkner J, Pawlowski N, Nichol A. Rasa: Open source language understanding and dialogue management. arXiv 2017. preprint arXiv:1712.05181

[61] Rojc M, Mlakar I, Kacic Z. The TTS-driven affective embodied conversational agent EVA, based on a novel conversational behavior generation algorithm. Engineering Applications of Artificial Intelligence. 2017;**57**:80-104

[62] Mlakar I, Smrke U. Clinical Study to Assess the Outcomes of a Patient-Centred Survivorship Care Plan Enhanced with Big Data and Artificial Intelligence Technologies. 2021. Available online: https://www.isrctn.com/ISRCTN97617326 [Accessed: June 19, 2021]

[63] Schoeffler M, Bartoschek S, Stöter FR, Roess M, Westphal S, Edler B, et al. Web MUSHRA—A comprehensive framework for web-based listening tests. Journal of Open Research Software. 2016:6. DOI: 10.5334/jors.187

[64] H2020 Project PERSIST. Available online: https://projectpersist.com/ [Accessed: May 31, 2020]

[65] Sofer C, Dotsch R, Wigboldus DH, Todorov A. What is typical is good: The influence of face typicality on perceived trustworthiness. Psychological Science. 2015;**26**(1):39-47

[66] Singh A, Ramasubramanian K, Shivam S. Introduction to Microsoft Bot, RASA, and Google Dialogflow. In: Building an Enterprise Chatbot: Work with Protected Enterprise Data Using Open Source Frameworks. Apress. Berkeley, CA, USA: Springer; 2019. pp. 281-302

[67] Ciechanowski L, Przegalinska A, Magnuski M, Gloor P. In the shades of the uncanny valley: An experimental study of human–chatbot interaction. Future Generation Computer Systems. 2018;**92**:539-548

Chapter 5

Towards Children-Centred Trustworthy Conversational Agents

Marina Escobar-Planas, Vicky Charisi, Isabelle Hupont, Carlos-D Martínez-Hinarejos and Emilia Gómez

Abstract

Conversational agents (CAs) have been increasingly used in various domains, including education, health and entertainment. One of the growing areas of research is the use of CAs with children. However, the development and deployment of CAs for children come with many specific challenges and ethical and social responsibility concerns. This chapter aims to review the related work on CAs and children, point out the most popular topics and identify opportunities and risks. We also present our proposal for ethical guidelines on the development of trustworthy artificial intelligence (AI), which provide a framework for the ethical design and deployment of CAs with children. The chapter highlights, among other principles, the importance of transparency and inclusivity to safeguard user rights in AI technologies. Additionally, we present the adaptation of previous AI ethical guidelines to the specific case of CAs and children, highlighting the importance of data protection and human agency. Finally, the application of ethical guidelines to the design of a conversational agent is presented, serving as an example of how these guidelines can be integrated into the development process of these systems. Ethical principles should guide the research and development of CAs for children to enhance their learning and social development.

Keywords: chatbots, children, conversational agents, trustworthy AI, ethics

1. Introduction

Conversational agents (CAs) are computer programs designed to engage in a conversation with humans through voice or text-based interactions [1]. Nowadays, their availability and popularity are dramatically increasing. They start to be embedded in a wide range of devices used on a daily basis and in ubiquitous ways: mobile phones, smart cars [2], home devices and even social robots and toys [3].

Very recent improvements in this technology are driven by ground-breaking artificial intelligence (AI) techniques [4, 5]. They are allowing for a brand new generation of CAs, such as ChatGPT [6] and GPT-4 [7], which have demonstrated unprecedented levels of autonomy and natural language processing capabilities. These systems have the potential to transform the way humans interact with computers and machines, offering simple and intuitive interfaces for a wide range of applications, from customer service [8] to entertainment [9], personal assistance [10], healthcare [11] or education [12]. However, this rapid progress also raises important ethical concerns, particularly when it comes to vulnerable populations such as children [13]. As such, there is a growing need for research on the ethical challenges to be tackled when it comes to the development and deployment of CAs that are conceived or that can potentially be used by children, to ensure that these systems are designed with the best interests of children in mind.

The goal of this chapter is to explore the unique considerations involved in the development and deployment of conversational agents for children. While there is a growing body of research on the ethical development of artificial intelligence in general, much of this work has yet to be fully applied to the specific challenges and opportunities of CAs for children. Our aim is to highlight the key ethical principles and best practices that should guide the design and implementation of these systems, with a focus on promoting safety, privacy and well-being for young users.

To achieve this goal, in Section 2, we begin by reviewing the related work on CAs and children, highlighting the most popular topics, such as educational and health applications. We also identify the opportunities and risks associated with the use of CAs for children. Next, we present ethical guidelines on the development of trustworthy AI, which provide a foundation for the ethical design and deployment of CAs with children. In Section 3, we present a case study that demonstrates the impact of CAs on children's learning and social development. Finally, in Section 4, we present an adaptation of the AI ethical guidelines to the specific case of CAs and children, highlighting the importance of data protection and human agency. We then demonstrate how these principles can be put into practice, offering a concrete example of how to design and deploy a CA that is both effective and ethical.

By the end of this chapter, readers will have a comprehensive understanding of the unique ethical challenges and opportunities presented by CAs for children, as well as a set of best practices for designing and implementing these systems in a responsible and ethical way.

2. Related work

2.1 Conversational agents

There are various terms used to describe conversational agents (CAs), such as dialog systems, virtual assistants and chatbots. These are all names for computer programs that allow people to interact with them through conversation [1] by using speech, text, or multimodal input/output. They have become popular in recent years and can be found on many devices such as smart speakers or cars. Traditional CAs typically consist of five main modules (**Figure 1**):

1. Automatic speech recognition (ASR): this module converts speech input into text, allowing the CA to understand and process what the user is saying.

2. Natural language understanding (NLU): this module takes the text generated by the ASR module and extracts the meaning behind it. The NLU module uses techniques such as semantic parsing to extract relevant information from the input.

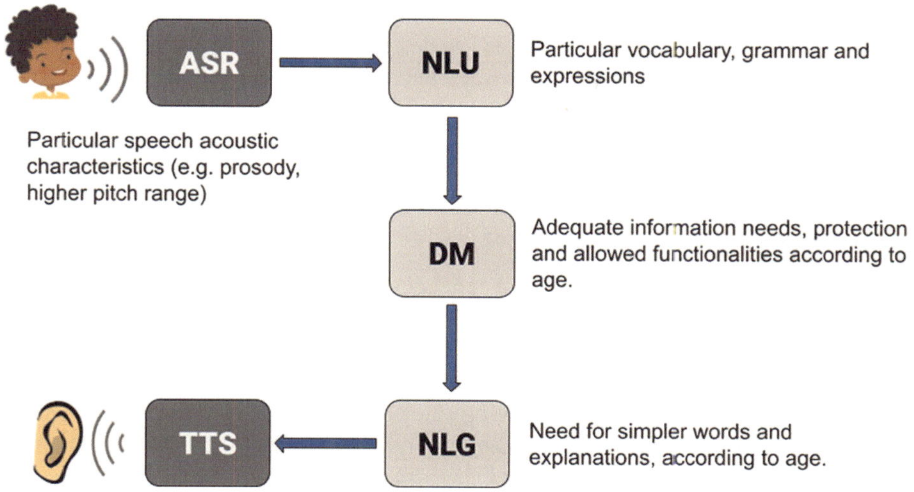

Figure 1.
Modules typically included in a conversational agent: ASR (automatic speech recognition), NLU (natural language understanding), DM (dialog management), NLG (natural language generation) and TTS (text-to-speech). The text close to each module summarizes the specific challenges that need to be tackled for their use with children.

3. Dialog management (DM): this module is responsible for managing the flow of the conversation and determining the appropriate response to the user's input based on the NLU module's understanding. It is usually based on identifying user intentions from this input to perform the needed actions and provide an answer to the user.

4. Natural language generation (NLG): this module takes the information obtained by the NLU and the answer type chosen by the DM module and generates a response in natural language for the user.

5. Text-to-speech (TTS): this module converts the text generated by the NLG module into speech output, allowing the CA to speak to the user in a natural way.

These modules work together to understand and respond to user input in a natural way. At present, many CAs are using advanced machine-learning methods to enhance their performance in one or more of these modules. For instance, BERT models are used to improve NLU [14], Reinforcement learning is used to improve DM [15] and ChatGPT utilizes deep neural networks to perform functions of NLU, DM and NLG in a unified manner.

CAs could be general, that is, without a specific purpose, such as the nowadays hugely popular ChatGPT [6] system. Alternatively, CAs could be intended to perform a quite specific mission, which is known as task-oriented CAs. This task-oriented nature makes CAs dependent on the corresponding use case they are developed for. For example, a CA that is devoted to home automation control expects interactions very different to a CA that is embedded in a social robot devoted to information tasks and guidance in commercial environments. The first system expects direct commands and it would ask for very few clarifications, whereas the other system will communicate in a more human-like form in order to establish a trusted linkage with human users.

Our focus in this chapter is on task-oriented CAs [16], which include tasks such as clothing selection [17], flight booking [18] and driving assistance [19]. CAs designed to assist users in completing specific tasks have become increasingly popular across various industries, such as e-commerce (e.g., purchasing), customer service (e.g., answering frequently asked questions) and healthcare (e.g., scheduling appointments). Some of these tasks are specifically designed for children, such as science learning [20], but CAs intended for adult-centred tasks are also utilized by children [21]. This may be due to the increasing accessibility and ease of use of these technologies for children.

2.2 Conversational agents and children

CAs are widely used devices, but it is important to consider their accessibility and popularity among children, as even young ones can interact with them through voice commands. In this section, we first present an overview of the literature in the field of conversational agents and children, and then an in-depth analysis of challenges, risks and opportunities identified in the related work.

2.2.1 Bibliometric study

We performed a bibliometric analysis to collect insights into how the research community has tackled the study of children and CAs. For that purpose, we used the *bibliometrix* tool [22] that allows extracting, among others, the most frequent keywords, clusters and co-occurrences of terms from a corpus of research papers. We collected the corpus of relevant papers from both Web of Science[1] and Scopus[2] as a result of the following search query over the papers' title, abstract and author keywords:

((“child” OR “children”) AND (“conversational agent” OR “conversational AI” OR “dialogue system” OR “dialogue systems” OR “chatbot” OR “chatbots” OR “virtual assistant” OR “home assistant” OR “voice assistant”))

We compiled a total of 440 papers, published from 2000 to 2022. Of the 440 papers, about 54% of them have been written in the period 2020–2022, with a clear increasing trend over the years, which is particularly remarkable since 2015 (from 7 papers per year in 2015 to 83 in 2022). This is probably a consequence of the recent popularity and market availability at a relatively low cost of conversational agents and assistants (e.g. Amazon's Alexa, Apple's Siri) and the emergence of large language models such as chatGPT [6].

Interestingly, when adding the terms *“social robot” OR “robot interaction”* to the search query, the number of papers increases from 440 to 2580, meaning that the research community has generally put a great effort into studying the effects of embodiment and the non-verbal side of interaction (e.g. through gestures, gaze or facial expressions).

For the verbal communication side, many studies on human-robot interaction with children actually rely on *Wizard-of-Oz* experimental settings [23, 24]. However, the full automation of voice interaction is already a reality and will be further enhanced by the aforementioned large language models in the near future. Nevertheless, careful attention has yet to be paid to their—still under-explored—maturity, moral capabilities [25] and trustworthiness when it comes to interacting with children.

---

1 Web of Science: https://www.webofscience.com/

2 Scopus: https://www.scopus.com/

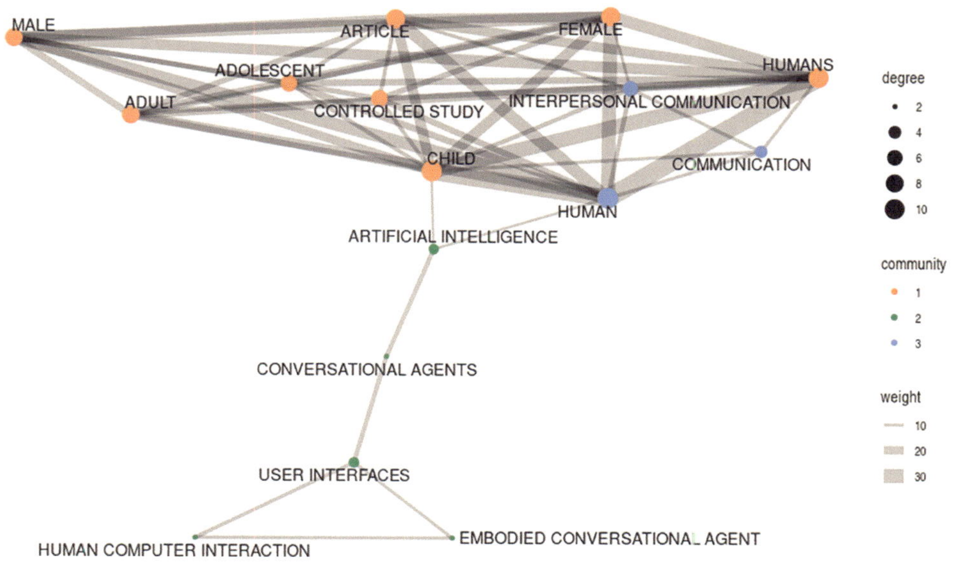

Figure 2.
Results from our bibliometric analysis of 440 papers focusing on CAs and children: Most frequent terms, cluster and co-occurrence network. The thicker the link, the more weight the co-occurrence of words has. The size of the nodes indicates the frequency of the keyword (the larger the radius, the greater the use) and their color (red, blue or green) denotes which cluster they belong to.

Figure 2 shows the results of the keyword frequency, co-occurrence and clustering analysis carried out over the 440 papers related to children and CAs. The bibliometric algorithm identifies three main clusters of terms (nodes). The most populated cluster and the one with more co-occurrences of keywords, is the one represented with red nodes, having the term "controlled study" at the centre, strongly linked to "humans" and demographic aspects including "male", "adult", "adolescent", "female" and "child". This cluster is therefore likely to be related to a large amount of literature presenting in-lab (controlled) experiments, where "adults" frequently act as guardians. The second cluster, with blue nodes and very close to (almost contained in) the first cluster, presents fewer but highly influential keywords, including "communication" and "interpersonal communication", which translates into studies analyzing the communicative—rather than or in addition to demographic—aspects of child-CA interaction in "controlled studies". The third cluster, with green nodes, appears clearly separated from the other two, which is an interesting finding as it contains more technical words such as "artificial intelligence", "user interfaces" and "(embodied) conversational agents". This decoupling of studies focusing on behavioral analyses from the more technical ones might translate into a gap to be bridged in the research community: fostering multidisciplinary collaborations between social scientists and AI/human-machine-interaction researchers, designers and engineers. This is of particular importance in this new era where large language models will pave the way towards fully automated (and less "controlled") human-CA interaction.

2.2.2 Risks and opportunities for children

Recent studies show that children actively use and explore CAs in home settings more than adults [21, 26, 27]. Thus, it is essential to consider the potential benefits and risks that these devices present for children when developing and implementing them.

In this context, we present and expand upon previous work [13, 28] regarding the risks, challenges and opportunities associated with these devices.

According to research, CAs bring several benefits to children, including:

- Engaging learning: by assisting in information search [29, 30], language learning [31] and school material learning [32, 33].

- Supporting health: by helping record treatments and track certain diseases [34] and chatting to reduce and control depression and anxiety [35].

- Enhancing accessibility: by enabling children with limited writing abilities (including toddlers), dyslexia or physical disabilities to interact with computers [36, 37].

- Improving social behavior: by promoting the use of persuasive strategies in games [38] and helping autistic children with social skills [39, 40].

However, as CAs are designed for a general population, their interactions with children may be affected by the unique characteristics and needs of children [41, 42]. Their language and communication abilities, as well as their particular rights, can pose challenges for the various components of the CA (**Figure 1**). The challenges of child-robot interaction have stimulated extensive research aimed at mitigating potential risks. Researchers have identified the following risks in the literature:

- Exclusion: children's different speech and understanding can lead to exclusion. Researchers have been working to enhance CA's performance for children, such as developing speech identification for babies [43] and identifying strategies for when the system does not understand a child [44].

- Over-trust: children may view CAs as friends [45, 46], which can lead to data disclosure risks. Transparent information, as demonstrated by Straten et al. [47], helps to reduce over-reliance on the system.

- Gender bias: chatbots are often designed with gender-specific cues and explicitly or implicitly conveyed as female [48].

- Unanticipated risks: the novelty of CAs makes it challenging to anticipate and prevent future problems, such as bullying experienced by girls named "Alexa" [49] due to Amazon's use of her name as an activation command for their device. Regular evaluation of these systems can help mitigate unanticipated risks.

Given those upcoming risks, guidelines are necessary to ensure the development of trustworthy CAs. These guidelines should take advantage of the benefits of these devices while assessing and minimizing potential risks and harm they may cause.

2.3 Trustworthy artificial intelligence

Several organizations worldwide have devoted efforts in the recent years to reflect on the ethical impact of artificial intelligence (AI) systems. The main goal of initiatives on AI and ethics is to raise awareness about the ethical considerations related to these systems, deepen our understanding of them and minimize the potential risks of AI while maximizing their benefits.

For example, the high level expert group (HLEG) of the European Commission developed the ethical guidelines for trustworthy AI [50] having in mind the respect for fundamental rights in various contexts where AI systems are used. These guidelines put forward a set of seven key requirements that AI systems should meet in order to be deemed trustworthy, which are:

1. Human agency and oversight: AI systems should respect human autonomy and decision-making and should be overseen by humans.

2. Technical robustness and safety: AI systems should be accurate, reliable and safe and should adopt a preventative approach to risks.

3. Privacy and data governance: AI systems should protect privacy and have legitimate access to data.

4. Transparency: AI systems should have clear documentation and inform users about their decisions, capabilities and limitations.

5. Diversity, non-discrimination and fairness: AI systems should promote inclusion throughout their entire life cycle.

6. Societal and environmental well-being: AI systems should benefit society and the world.

7. Accountability: AI systems should have mechanisms in place to ensure responsibility for their development, deployment and use.

These ethical guidelines are complemented by an assessment list for trustworthy AI (ALTAI) [51], which is designed as a practical tool to help organizations self-assess the trustworthiness of their AI systems. ALTAI is a list of 69 self-evaluation questions grouped into the aforementioned seven requirements. Although these ethics guidelines are designed for general populations, it refers to children as a relevant vulnerable population, and it states the need to pay particular attention to them.

With a focus on children, UNICEF developed a policy guidance [52] that aims to raise awareness of children's rights in the context of AI systems. The guidance is based on nine requirements, including: (1) supporting children's development and well-being, (2) promoting inclusiveness for children, (3) prioritizing fairness and avoiding discrimination for children, (4) protecting children's data and privacy, (5) ensuring safety for children, (6) providing transparency, explainability and accountability for children, (7) empowering knowledge of AI and children's rights, (8) preparing children for present and future AI developments and (9) creating an enabling environment.

Previous research [13] has shown that most of the requirements in UNICEF's policy guidance for AI and children align with HLEG ALTAI (**Table 1**), except for requirement 8, which focuses on educational policies, of special relevance for children and broadly addressed by the HLEG in the context of jobs and skills. Despite this alignment, the two guidelines differ in their focus: while UNICEF's ones include policy considerations, HLEG places an emphasis on the development and evaluation of AI systems.

UNICEF AI for children	**HLEG ALTAI**						
	1	2	3	4	5	6	7
1. Supporting children's development and well-being		xx	xx		xx	xxx	
2. Promoting inclusiveness for children					xxx		
3. Prioritizing fairness and avoiding discrimination for children					xxx		
4. Protecting children's data and privacy	xx		xxx				
5. Ensuring safety for children	xx	xxx				xx	
6. Providing transparency, explainability and accountability for children	xxx	xx		xxx	xx		xxx
7. Empowering knowledge of AI and children's rights				x		xxx	
8. Preparing children for present and future AI developments	x			x		x	
9. Creating an enabling environment	xx				xx	xx	

This table shows the correspondence between HLEG ALTAI's seven requirements (1) Human agency and oversight, (2) Technical robustness and safety, (3) Privacy and data governance, (4) Transparency, (5) Diversity, non-discrimination, and fairness, (6) Societal and environmental well-being, and (7) Accountability, and the corresponding requirements in UNICEF's AI for children guidelines (rows). The cells are marked with x, xx, or xxx to indicate the degree of correspondence between the related requirements, with low correspondence indicated by x, mid-level correspondence indicated by xx, and high correspondence indicated by xxx.

Table 1.
Mapping between HLEG ALTAI and UNICEF AI for children requirements.

A recent report by the Joint Research Centre of the European Commission [28] recognized the need for a connection between existing research in the area of AI and children's rights and the current policy initiatives and needs, proposing an integrated agenda for research and policy. In order to do that, the authors conducted a series of workshops with policymakers, researchers and children to gain insights into the interplay between different stakeholders, to connect scientific evidence with policymaking and to change the focus from the identification of ethical guidelines towards the definition of methods for practical future AI implementations. The report highlights the need for strategic and systemic choices to develop AI-based services that limit the use of AI to tasks that serve a valuable purpose. It emphasizes the importance of minimizing the environmental impact of AI technology, particularly in reducing CO_2 emissions from data centres. Developers must ensure that AI technology is child-friendly and free of discriminatory biases, while children must have control over their personal data. Transparency, explainability and accountability are critical to empowering young users of AI technology. The report also stresses the need for further research to understand how agency is developed in children when interacting with AI-based systems.

3. A scenario from the field of child-robot interaction

In this section, we present current scientific evidence regarding the impact of conversational agents on children. The objective is to emphasize the importance of ethical guidelines, as outlined in Section 2.3, and to highlight the need for their adaptation in the context of CAs and children.

3.1 Motivation and rationale

As discussed in previous sections, conversational agents are present in various contexts and embodiments. For instance, a written interaction with an open-domain chatbot on a computer differs significantly from a voice interaction with a driving assistant in a car. We acknowledge that the embodiment and context of the conversational agent impact the user's perception and behavior, beyond the dialog. However, given the diverse range of conversational agents available and the findings of our literature review in Section 2.2.1, which showed the popularity of child-robot interaction studies, we have chosen to present a use case of a social robot in a controlled educational context.

To illustrate the potential impact of CAs on children, this use case reflects current research on social robots and builds upon previous research with educational focus [53–55]. This aligns with procedures suggested by UNICEF and provides a relevant scientific context of application of CAs.

Specifically, we discuss a large-scale experimental study with hybrid small-group settings, where we investigated the effects of certain robot behaviors on children's problem-solving processes, social dynamics and perceptions of the robot. This experimental study serves as a starting point for identifying emerging issues that could apply to other conversational agents. In the following subsections, we provide a brief overview of the methodology and results of the study. For a more detailed description, we direct the reader to ref. [56, 57].

3.2 Methodology

For this experimental study, 84 children (5–8 years old) were recruited and divided into pairs to collaborate with a conversational agent embodied in a table-top social robot in order to perform a Tower of Hanoi problem-solving task (see **Figure 3**). The study was structured as follows: In the preliminary session, children were introduced to the Hanoi tower game. In the robot intervention session, they collaborated with the robot to solve the task. The interview session involved a semi-structured interview and a picture task to capture children's perceptions the robot.

We manipulated two variables of the CA: its cognitive reliability and its expressivity, which are described as follows:

- In terms of cognitive reliability, the dialog management (DM) module allowed the CA to provide either optimal or non-optimal suggestions to solve the problem-solving task, depending on the selected behavior.

- In terms of expressive behavior, the natural language generation (NLG) module was configured to create two different behaviors: an expressive version of the system, which employed more emotive and engaging phrasings like "What do you think super-team? Do you feel like playing again?"; and a neutral version that used phrasings like "Would you like to repeat the game?". In addition, these verbal expressions were coupled with different configurations of the text-to-speech (TTS) module, as outlined in **Table 2**, to control the expressivity of the robot's

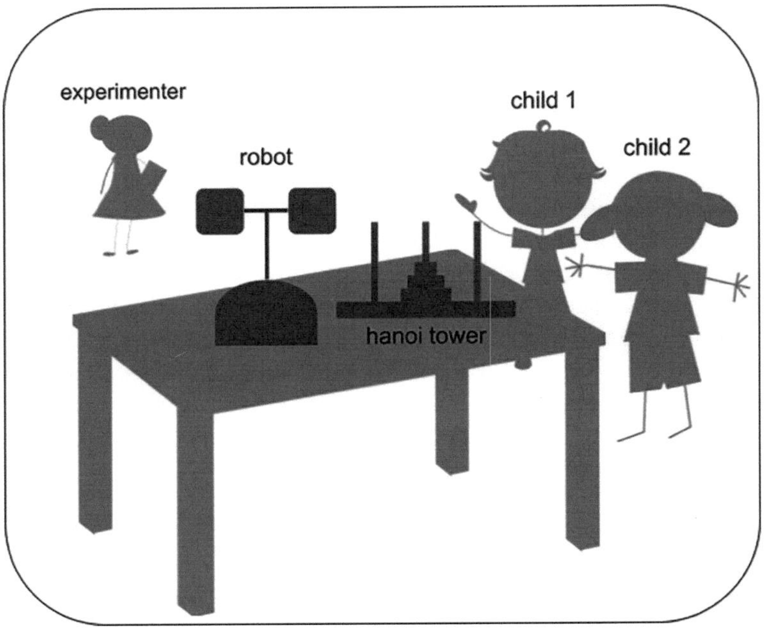

Figure 3.
Experimental setup for studying collaborative problem-solving between two children and a table-top social robot using the tower of Hanoi logic task.

sentences. While in the neutral condition, the CA has used all the time the *n*eutral configuration and the expressive condition changed between *original, calm* and *happy* configurations. We selected two different features, pitch and speed, and empirically set those parameters during the design phase of the study [56].

In the study, we measured the task performance and group dynamics during the interaction. We also interviewed children to understand both the influence of the different behaviors and children's perception of the system.

3.3 Results

We summarize here the results of the study, with a focus on the link between the mentioned ethical guidelines and the communication features employed by the social robot during interactions.

The analysis of the recorded interviews showed that children generally find the social robot to be friendly, able to perceive them and not able to harm anyone (over 90% positive answers). The thematic analysis also revealed two important concepts:

Expressions	Mood	Speed	Pitch
Neutral	Serious	80%	−3
Expressive	Original	100%	0
	Calm	85%	0
	Happy	120%	+7

Table 2.
TTS features in both expressive and neutral conditions.

perceived autonomy, meaning children's perceptions about the robot's ability to act on its own even when they know the robot has to follow its programming, and shared responsibility, which involves working together towards a common goal, fostering a sense of understanding and teamwork among the group members.

Additionally, children were affected by the different robot behaviors, including the ones related to the CAs described above.

- **Task performance.** A low performance by the robot in the tower of Hanoi task led to a decrease in the overall task performance and changed the dynamics of the group, as one child often took the lead. Children who perceived the robot's low performance recognized their own help to the robot, leading to a better-perceived help relationship between the child and the robot. This collaboration, combined with a probable lack of confidence in solving the task, resulted in children asking for the robot's assistance more quickly. On the other hand, children in the condition of a cognitive reliable robot who experienced high performance from the robot were more self-assured in solving the task and took longer to request help, even though they perceived more help from the robot.

- **Expressivity.** A high level of expressivity displayed by the robot during the interaction led to a stronger perception of the robot as a friend and less as a machine. Children also viewed the robot as more capable of assisting them with their homework, even when its task performance was low, emphasizing the social support it provided rather than its accuracy in the homework task. The researchers noted that children were able to distinguish the behavior of the expressive robot when compared with the non-expressive one; however, during the interaction, there was a tendency to view the robot as being expressive even when it was not.

Overall, this experiment provides valuable insights into children's perceptions of an embodied CA and the impact of the DM (task performance) and DM + NLG (expressivity) modules. The results indicate that children had a positive opinion about the CA, but a low task performance affects children's social dynamics and a high expressivity increases its perception as a friend and children's aim to do homework with it. These examples help us to illustrate the impact CAs have on children and help to understand the role of CAs in children's lives.

4. Towards trustworthy conversational agents for children

In previous sections, we have demonstrated the potential impact of CAs on children underscoring the need for developing trustworthy CAs that are suitable for them. Previous research on AI and children [28] has acknowledged the need to move from the identification of ethical guidelines to practical implementation. Building upon our experience in the study described in Section 3, in this section, we introduce an adaptation of ethical guidelines for CAs that consider children as potential users, along with their implementation in a practical system design.

4.1 Adaptation of ethical guidelines to CAs and children

A recent study aimed to adapt ethical guidelines for AI systems to the specific case of CAs and children [13]. A team of four experts in computer science, AI ethics and children's rights evaluated each item of ALTAI (introduced in Section 2.3) in terms of relevance and particular considerations for CAs and children. A Delphi method approach [58] was followed to perform a risk level analysis [59] as follows. The experts rated each ALTAI item based on the likelihood of application and impact on children and CAs. Later on, the individual ratings were analysed to identify critical points and disagreements. Finally, the identified disagreements were discussed and resolved at an expert meeting in order to reach a consensus.

The quantitative analysis consisted of a risk level assessment, using the formula $Risk = Likelihood \times Impact$ to measure the level of risk of the different items and sections in ALTAI. A thematic analysis was also carried out on the annotated comments provided by the experts. The study's main findings are summarized below.

- **Stakeholders' involvement:** The experts emphasized the importance of involving children, teachers and parents as stakeholders in the design, use and testing of the system. Some experts emphasized the need for multiple stakeholders to be involved and for stakeholders to be taught to help oversee the system. All experts agreed that children's involvement should be done in a meaningful and entertaining way, as they should not be put in job-like conditions.

- **Risk management:** The importance of risk management for children, as a vulnerable population, was stressed by the experts during the CA development. They recommended implementing high measures for data storage to ensure privacy and security of personal data, avoiding access by third parties. To guarantee personal data privacy and security, the experts recommended implementing high measures for data storage and prevent access by third parties. They also suggested defining metrics and risk levels to monitor system performance and facilitate testing, evaluation and external audits. Allowing users to write reports about the system can also aid in identifying risks and errors. Finally, they emphasized the importance of transparency in addressing privacy concerns and minimizing children's data disclosure.

- **AI Awareness:** Experts underscored the importance of emphasizing the non-human aspect of CAs when designing them for children. This is important to minimize the attachment that children may form towards them and reduce their influence on the child. Maximizing the user's agency is also recommended to further reduce the impact. For example, providing multiple options when a recommendation is requested. To maintain transparency, constant access to the system's information, including its nature, functions and limitations, can be provided.

- **Age-appropriate behavior:** Inclusivity, as highlighted by the experts, is a vital aspect of children's education and development. Therefore, it is important to mitigate any technical difficulties that may arise when CAs interact with children or marginalized groups. In case of any breakdown, a reliable recovery strategy can help to continue the interaction. Additionally, guardians should take responsibility for children when they are using CAs. In the event of a problem, they should be contacted for assistance, and double consent mechanisms should be in place considering the guardian and the child. To enhance critical thinking and self-regulation, transparency can be applied by using a language that is adapted to the user's age.

- **Transparency.** According to the experts, transparency is an important factor that can help address several critical considerations mentioned before. The CA trustworthiness could be improved by providing information about the system's nature, privacy and limitations in a language that is appropriate for the user's age.

The findings of this study are largely consistent with previous work on ethical guidelines for AI [28, 51, 52] but provide a more in-depth perspective of the specific issues related to CAs and children. The risk level assessment revealed that although all the identified points are critical, some present higher risks than others (**Table 3**). Particularly, the concerns of children were rated as having a higher impact, while the likelihood of issues was higher for CAs, resulting in a higher overall risk for CAs. Based on these findings, "Privacy and data governance" and "Human agency and oversight" were identified as the two critical requirements that should be prioritized when developing CAs for children.

For more details on the methodology and results of this study, please refer to ref. [13]. The critical points identified in this study, which have been taken into account in subsequent work [60], were applied throughout an agent's design process as illustrated in the subsequent section.

4.2 Proposal of a system

In this section, we introduce a previous study [60] that applied the presented guidelines (Section 4.1) to the development of a conversational agent intended to create a list of games and toys according to user preferences.

The design of the interaction is limited to one user at a time, and the system can ask questions to provide the user with a list of interesting items to choose from. The system can also ask for data such as hobbies, idols and cost limits to fill the user's profile and determine the restrictions on the products that can be offered. The algorithm for the general CA is presented in Algorithm 1.

HLEG requirement	Risks to Children[a]	Risks to CAs[a]	Total Risk[b]
Human agency and oversight	**5.21***	**5.87***	**30.59***
Technical robustness and safety	3.38	4.10	13.85
Privacy and data governance	4.99	**6.96***	**34.75***
Transparency	4.56	**5.14***	**23.43***
Diversity, non-discrimination and fairness	4.14	**5.71***	**23.63***
Societal and environmental well-being	2.59	3.00	7.78
Accountability	3.10	4.93	15.28

[a] *Risk = Likelihood × Impact.*
[b] *Total Risk = Risk on Children × Risk on CAs.*
High risk values are marked with an (5-9 for Risks to Children and Risks to CAs, and 18-81 for the Total Risk).*

Table 3.
Risk assessment results based on expert evaluations of the likelihood and impact of questions in conversational agents and children.

The article proposed modifications consider children as possible users, including changes to the design, technology, interaction and post-interaction phases. These are summarized in **Table 4** and described in detail in the following paragraphs.

- Regarding the design phase, the "game recommender system" is redefined as a "games wish list" to improve children's interest. A list of relevant stakeholders is defined to gather ideas on how to adapt the system to a children-friendly version. Stakeholders are involved in defining age ranges to be considered, as well as, in time limits for interaction, identification of positive game properties and evaluation of the system.

Algorithm 1: Pseudo-code for the proposed CA.

```
1  Welcome message
2  Ask if the user wants a specific item or a suggestion
3  if user input request to end then
4  |   Start Goodbye
5  |   Display list
6  else if user input request a specific item then
7  |   Add item to the list
8  else if user input request a suggestion then
9  |   Check if the profile information is enough for suggestions
10 |   while information not enough do
11 |   |   Ask information to the user to fill the profile
12 |   Provide suggestions
13 |   Go to line 3 (Ask to add any items to the list)
14 Ask information to the user to fill the profile
15 Go to line 3 (starts the interaction again)
```

- Regarding technology, the article suggests minimizing the use of standard search engines and black box approaches in the DM and NLG modules. A safety check module could improve the system's reliability. The storage of personal data is minimized, and all data is stored in a secure server to prevent access by third parties. The ASR module is chosen with a good understanding rate for children and vulnerable populations to maximize the system's inclusivity.

- Regarding the interaction phase, modifications are proposed, including guessing the user's age range, using adapted vocabulary, splitting the welcome message into a guardian and a child welcome message and informing the user about the system's non-human nature, confidentiality and algorithmic decisions. The system should incorporate a control mechanism when accessing online search engines, filter out non-adequate items for children, promote neutral games, or always suggest some toys associated with another gender, display at least three varied suggestions and have a good recovery strategy.

- Regarding the post-interaction phase, audits with access to the system's metrics and reports will help identify critical and not critical problems.

General	Specific	Particular measures	Line of code
Stakeholders' involvement	Through the CA lifecycle	• Define features (e.g. age ranges and max interaction time).	-
		• Consult stakeholders throughout design, implementation and evaluation.	-
Risk management	Privacy measures	• Minimize the personal data to be stored.	11, 14
		• Do not allow additional usage/transfer of stored data.	-
	Security measures	• Reduce standard black boxes and search engine usage in DM and NLG.	-
		• Incorporate a control mechanism for online search.	12
		• Define trigger keywords for guardian involvement (e.g. weapons and sex).	2, 6, 11, 14
		• Store data in a safe server with cybersecurity measures.	-
		• Define metrics for risk management, e.g. time spent with the user, times the system calls the guardian.	-
	Facilitate reports	• After the interaction, gather feedback from children and guardians.	After 5
		• Offer accessible error reporting and mention it in the welcome message.	1, after 5
AI awareness	Access to information	• Include concise relevant CA information in the welcome message and pointers to additional details.	1
		• Inform about the system's not-human and not-feeling nature.	1
		• Inform about CA's confidentiality and algorithmic decisions.	1
	Influence	Configure the system to display at least three suggestions.	12
Age approp. Behavior	Guardians	• Split welcome message: guardian and child. Need two consents.	1
		• Invoke guardians in security issues (e.g. dangerous requests or persistent breakdowns).	2, 11, 14
	Education and self-development	• Define toys-classification to benefit children's development. Consider them for suggestions.	12
		• Consider gender bias in recommended items.	12
		• Control and communicate the time spent on the interaction.	Before 14
	Inclusivity	• Guess/ask for the user's age at the beginning of the interaction.	1
		• Define functionality as a "wish list" if a child is recognized.	1
		• Adapt the list of recommended items to age.	12
		• Adapt the vocabulary of the interaction to age.	1, 2, 4, 5, 11, 12, 14
		• Choose an inclusive ASR module.	3, 6, 8, 11, 14
		• Minimize neutral responses in breakdowns.	3, 6, 8, 11, 14

Table 4.
Recommendations to the design of a CA that generates a list of preferred toys/games for children.

Overall, the proposed modifications aim to improve the system's inclusivity and reliability, promote critical thinking and decrease overtrust and data disclosure.

5. Conclusions and challenges for future research

The field of conversational agents with children has made significant strides in recent years. Although extensive research has been conducted on the ethical development of artificial intelligence (AI), in general, there has been a growing emphasis on employing AI in interactions with children. Such work has emphasized the need for systems that enhance opportunities while mitigating risks.

However, the development and deployment of conversational agents for children come with specific challenges and ethical and social responsibility concerns. This chapter is dedicated to exploring the particular considerations conversational agents should have when interacting with children.

The chapter reviewed related research on conversational agents and children, identifying popular topics as well as opportunities and risks. Worldwide ethical guidelines on the development of trustworthy AI were presented as a framework for the ethical design and deployment of conversational agents with children. These guidelines emphasize the importance of the protection of user rights in the development and deployment of AI technologies. Additionally, a case study was presented that demonstrated the significant impact conversational agents can have on children's learning and social development.

An adaptation of previous AI ethical guidelines to the specific case of conversational agents and children was also presented, highlighting the importance of data protection and human agency. The application of ethical guidelines to the design of a conversational agent presented in this chapter served as an example of how these guidelines can be integrated into the development process of these systems.

It is important to note that the variety of conversational agent systems requires personalized study and application of these guidelines for each case. Furthermore, even state-of-the-art technology may not be able to address some proposed considerations, such as ASR modules that cannot understand little children's speech in less popular languages. Therefore, researchers in this area should continue to strive towards achieving new breakthroughs that enable the development of more ethically sound devices for the benefit of future generations.

In summary, the development and deployment of conversational agents with children require a careful balance between innovation and ethical responsibility. Ethical principles should guide research and development, and systems should be designed with the safety, privacy and well-being of children in mind. By doing so, conversational agents have the potential to be a powerful tool for enhancing children's learning and social development.

Acknowledgements

This work was carried out with the support of the Joint Research Centre of the European Commission in the framework of the Collaborative Doctoral Partnership Agreement No.35500.

Conflict of interest

The authors declare no conflict of interest.

Author details

Marina Escobar-Planas[1,2]*, Vicky Charisi[2], Isabelle Hupont[2], Carlos-D Martínez-Hinarejos[1] and Emilia Gómez[2]

1 Universitat Politècnica de València, Valencia, Spain

2 European Commission, Joint Research Center, Seville, Spain

*Address all correspondence to: marescplajob@gmail.com

References

[1] McTear M. Conversational ai: Dialogue systems, conversational agents, and chatbots. Synthesis Lectures on Human Language Technologies. 2020; **13**(3):1-251

[2] Lee K, M, Moon Y, Park I, Lee J-g. Voice orientation of conversational interfaces in vehicles. Behaviour & Information Technology. 2023:1-12. DOI: 10.1080/0144929X.2023.2166870

[3] Garg R, Cui H, Seligson S, Zhang B, Porcheron M, Clark L, et al. The last decade of hci research on children and voice-based conversational agents. In: Proceedings of the 2022 CHI Conference on Human Factors in Computing Systems. Vol. 1–19. New York, NY, United States: Association for Computing Machinery; p. 2022

[4] Singh S, Beniwal H. A survey on near-human conversational agents. Journal of King Saud University-Computer and Information Sciences. 2022;**34**(10): 8852-8866

[5] Diederich S, Brendel AB, Morana S, Kolbe L. On the design of and interaction with conversational agents: An organizing and assessing review of human-computer interaction research. Journal of the Association for Information Systems. 2022;**23**(1): 96-138

[6] van Dis EAM, Bollen J, Zuidema W, van Rooij R, Bockting CL. Chatgpt: Five priorities for research. Nature. 2023; **614**(7947):224-226

[7] Bubeck S, Chandrasekaran V, Eldan R, Gehrke J, Horvitz E, Kamar E, et al. Sparks of artificial general intelligence: Early experiments with gpt-4. arXiv. 2023

[8] Ngai EWT, Lee MCM, Luo M, Chan PSL, Liang T. An intelligent knowledge-based chatbot for customer service. Electronic Commerce Research and Applications. 2021;**50**:101098

[9] Simpson J, Nalepka P, Kallen RW, Dras M, Reichle ED, Hosking SG, et al. Conversation dynamics in a multiplayer video game with knowledge asymmetry. Frontiers in Psychology. 2022;**13**: 1039431. DOI: 10.3389/fpsyg.2022.1039431

[10] Ansar S, A, Jaiswal K, Aggarwal S, Shukla S, Yadav J, Soni N. Smart home personal assistants: Fueled by natural language processor and blockchain technology. In: 2022 Second International Conference on Interdisciplinary Cyber Physical Systems (ICPS). IEEE; 2022. pp. 113-117

[11] Sawad AB, Narayan B, Alnefaie A, Maqbool A, Mckie I, Smith J, et al. A systematic review on healthcare artificial intelligent conversational agents for chronic conditions. Sensors. 2022;**22**(7): 2625

[12] Khosrawi-Rad B, Rinn H, Schlimbach R, Gebbing P, Yang X, Lattemann C, et al. Conversational agents in education–a systematic literature review. In: Proceedings of the 30th European Conference on Information Systems (ECIS). Timis, Oara, Romania (Forthcoming). 2022. Available from: https://aisel.aisnet.org/ecis2022/. ISBN: 978-1-958200-02-5

[13] Escobar-Planas M, Gómez E, Martínez-Hinarejos C-D. Guidelines to develop trustworthy conversational agents for children. 2022:342-360. Available from: https://sites.utu.fi/ethicomp2022/wp-content/uploads/sites/1104/2022/09/Ethicomp-2022-Proceed

ings_Corrected.pdf. ISBN: 978-951-29-8989-8

[14] Devlin J, Chang M-W, Lee K, Toutanova K. Bert: Pre-training of deep bidirectional transformers for language understanding. arXiv. 2018. DOI: 10.48550/arXiv.1810.04805

[15] Lipton Z, Li X, Gao J, Li L, Ahmed F, Deng L. Bbq-networks: Efficient exploration in deep reinforcement learning for task-oriented dialogue systems. In: Proceedings of the AAAI Conference on Artificial Intelligence. Vol. 32. Palo Alto, California USA: AAAI Press; 2018

[16] Gao J, Galley M, Li L. Neural approaches to conversational ai. In: The 41st International ACM SIGIR Conference on Research & Development in Information Retrieval. New York, NY, United States: Association for Computing Machinery; 2018. pp. 1371-1374

[17] Sousa RG, Ferreira PM, Costa PM, Azevedo P, Costeira JP, Santiago C, et al. Ifetch: Multimodal conversational agents for the online fashion marketplace. In: Proceedings of the 2nd ACM Multimedia Workshop on Multimodal Conversational AI. New York, NY, United States: Association for Computing Machinery; 2021. pp. 25-26

[18] Wei W, Le Q, Dai A, Li J. AirDialogue: An Environment for Goal-Oriented Dialogue Research. In: Proceedings of the 2018 Conference on Empirical Methods in Natural Language Processing. Brussels, Belgium: Association for Computational Linguistics; 2018. pp. 3844-3854. Available from: https://aclanthology.org/D18-1

[19] Eric M, Krishnan L, Charette F, Manning CD. Key-value retrieval networks for task-oriented dialogue. In: Proceedings of the 18th annual SIGdial meeting on discourse and dialogue. Saarbrücken, Germany: Association for computational linguistics; 2017. pp. 37-49. DOI: 10.18653/v1/W17-5506

[20] Xu Y, Vigil V, Bustamante AS, Warschauer M. "elinor's talking to me!": Integrating conversational ai into children's narrative science programming. In: Proceedings of the 2022 CHI Conference on Human Factors in Computing Systems. Vol. 1–16. New York, NY, United States: Association for Computing Machinery; p. 2022

[21] Garg R, Sengupta S. He is just like me: A study of the long-term use of smart speakers by parents and children. Proceedings of the ACM on Interactive, Mobile, Wearable and Ubiquitous Technologies. 2020;**4**(1):1-24

[22] Aria M, Cuccurullo C. Bibliometrix: An r-tool for comprehensive science mapping analysis. Journal of Informetrics. 2017;**11**(4):959-975

[23] Nasir J, Oppliger P, Bruno B, Dillenbourg P. Questioning wizard of oz: Effects of revealing the wizard behind the robot. In: 2022 31st IEEE International Conference on Robot and Human Interactive Communication (RO-MAN). IEEE; 2022. pp. 1385-1392

[24] Zou J, Gauthier S. Salvatore M Anzalone, David Cohen, and Dominique Archambault. A wizard of oz interface with qtrobot for facilitating the handwriting learning in children with dysgraphia and its usability evaluation. In: Computers Helping People with Special Needs: 18th International Conference, ICCHP-AAATE 2022, Lecco, Italy, July 11–15, 2022, Proceedings, Part II. Springer; 2022. pp. 219-225. Available from:

https://link.springer.com/book/10.1007/978-3-031-08645-8

[25] Ganguli D, Askell A, Schiefer N, Liao T, Lukošiūtė K, Chen A, et al. The capacity for moral self-correction in large language models. arXiv. 2023. Available from: https://arxiv.org/abs/2302.07459

[26] Sciuto A, Saini A, Forlizzi J, Hong JI. "Hey alexa, what's up?" a mixed-methods studies of in-home conversational agent usage. In: Proceedings of the 2018 Designing Interactive Systems Conference. New York, NY, United States: Association for Computing Machinery; 2018. pp. 857-868. DOI: 10.1145/3196709

[27] Lovato SB, Piper AM, Wartella EA. Hey google, do unicorns exist? Conversational agents as a path to answers to children's questions. In: Proceedings of the 18th ACM International Conference on Interaction Design and Children. New York, NY, United States: Association for Computing Machinery; 2019. pp. 301-313. DOI: 10.1145/3311927

[28] Charisi V, Chaudron S, Di Gioia R, Vuorikari R, Planas ME, Sanchez MJI, et al. Artificial Intelligence and the Rights of the Child: Towards an Integrated Agenda for Research and Policy. Technical report, Joint Research Centre of the European Commission. (JRC127564, Joint Research Centre (Seville site)); 2022

[29] Landoni M, Murgia E, Huibers T, Pera MS. You've got a friend in me: Children and search agents. In: Adjunct Publication of the 28th ACM Conference on User Modeling, Adaptation and Personalization. New York, NY, United States: Association for Computing Machinery; 2020. pp. 89-94. DOI: 10.1145/3386392

[30] Downs B, French T, Wright KL, Pera MS, Kennington C, Fails JA. Children and search tools: Evaluation remains unclear. In: KidRec Workshop co-Located with ACM IDC 2019. 2019

[31] Gilani SN, Traum D, Merla A, Hee E, Walker Z, Manini B, et al. Multimodal dialogue management for multiparty interaction with infants. In: Proceedings of the 20th ACM International Conference on Multimodal Interaction. New York, NY, United States: Association for Computing Machinery; 2018. pp. 5-13. DOI: 10.1145/3242969

[32] Law E, Ravari PB, Chhibber N, Kulic D, Lin S, Pantasdo KD, et al. Curiosity notebook: A platform for learning by teaching conversational agents. In: Extended Abstracts of the 2020 CHI Conference on Human Factors in Computing Systems. New York, NY, United States; Association for Computing Machinery; 2020. pp. 1-9. DOI: 10.1145/3334480

[33] Xu Y, Warschauer M. "Elinor is talking to me on the screen!" integrating conversational agents into children's television programming. In: Extended Abstracts of the 2020 CHI Conference on Human Factors in Computing Systems. New York ,NY, United States; Association for Computing Machinery: 2020. pp. 1-9. DOI: 10.1145/3334480

[34] Sezgin E, Noritz G, Elek A, Conkol K, Rust S, Bailey M, et al. Capturing at-home health and care information for children with medical complexity using voice interactive technologies: Multi-stakeholder viewpoint. Journal of Medical Internet Research. 2020;**22**(2):e14202

[35] Fitzpatrick KK, Darcy A, Vierhile M. Delivering cognitive behavior therapy to

young adults with symptoms of depression and anxiety using a fully automated conversational agent (woebot): A randomized controlled trial. JMIR mental health. 2017;**4**(2):e7785

[36] Fabio Catania, Pietro Crovari, Eleonora Beccaluva, Giorgio De Luca, Erica Colombo, Nicola Bombaci, and Franca Garzotto. Boris: a spoken conversational agent for music production for people with motor disabilities. In *CHItaly 2021: 14th Biannual Conference of the Italian SIGCHI Chapter*, pages 1–5, 2021.

[37] Pradhan A, Mehta K, Findlater L. "Accessibility came by accident" use of voice-controlled intelligent personal assistants by people with disabilities. In: Proceedings of the 2018 CHI Conference on Human Factors in Computing Systems. New York, NY, United States: Association for Computing Machinery; pp. 1, 2018-13

[38] Fraser J, Papaioannou I, Lemon O. Spoken conversational ai in video games: Emotional dialogue management increases user engagement. In: Proceedings of the 18th International Conference on Intelligent Virtual Agents. New York, NY, United States; Association for Computing Machinery; 2018. pp. 179-184. DOI: 10.1145/3267851

[39] Ali MR, Razavi SZ, Langevin R, Al Mamun A, Kane B, Rawassizadeh R, et al. A virtual conversational agent for teens with autism spectrum disorder: Experimental results and design lessons. In: Proceedings of the 20th ACM International Conference on Intelligent Virtual Agents. New York, NY, United States: Association for Computing Machinery; 2020. pp. 1-8. DOI: 10.1145/3383652

[40] Zhang L, Weitlauf AS, Amat AZ, Swanson A, Warren ZE, Sarkar N. Assessing social communication and collaboration in autism spectrum disorder using intelligent collaborative virtual environments. Journal of Autism and Developmental Disorders. 2020;**50**: 199-211

[41] Narayanan S, Potamianos A. Creating conversational interfaces for children. IEEE Transactions on Speech and Audio Processing. 2002;**10**(2):65-78

[42] Kennedy J, Lemaignan S, Montassier C, Lavalade P, Irfan B, Papadopoulos F, et al. Child speech recognition in human-robot interaction: Evaluations and recommendations. In: Proceedings of the 2017 ACM/IEEE International Conference on Human-Robot Interaction. New York, NY, United States: Association for Computing Machinery; 2017. pp. 82-90. DOI: 10.1145/2909824.3020229

[43] Lavechin M, Bousbib R, Bredin H, Dupoux E, Cristia A. An open-source voice type classifier for child-centered daylong recordings. arXiv. 2020

[44] Røyneland K. "It Knows How to Not Understand us!" a Study on What the Concept of Robustness Entails in Design of Conversational Agents for Preschool Children [Thesis]. Oslo, Norway: University of Oslo; 2019

[45] Kahn Jr PH, Kanda T, Ishiguro H, Freier NG, Severson RL, Gill BT, et al. "Robovie, you'll have to go into the closet now": Children's social and moral relationships with a humanoid robot. Developmental Psychology. 2012;**48**(2): 303

[46] Druga S, Williams R, Breazeal C, Resnick M. Hey google is it ok if i eat you? initial explorations in child-agent interaction. In: Proceedings of the 2017 Conference on Interaction Design and

Children. New York, NY, United States: Association for Computing Machinery; 2017. pp. 595-600. DOI: 10.1145/3078072

[47] van Straten CL, Peter J, Kuhne R, Barco A. Transparency about a robot's lack of human psychological capacities: Effects on child-robot perception and relationship formation. ACM Transactions on Human-Robot Interaction (THRI). 2020;**9**(2):1-22

[48] Tolmeijer S, Zierau N, Janson A, Wahdatehagh JS, Leimeister JMM, Bernstein A. Female by default?–exploring the effect of voice assistant gender and pitch on trait and trust attribution. In: Extended Abstracts of the 2021 CHI Conference on Human Factors in Computing Systems. New York, NY, United States: Association for Computing Machinery; pp. 1, 2021-7. DOI: 10.1145/3411763

[49] Johns T. Parents of Children Called Alexa Challenge Amazon. BBC. Available from: https://www.bbc.com/news/technology-57680173

[50] HLEG AI. Ethics Guidelines for Trustworthy Ai. Brussels; 2019. Available from: https://digital-strategy.ec.europa.eu/en/library/ethics-guidelines-trustworthy-ai

[51] Ala-Pietilä P, Bonnet Y, Bergmann U, Bielikova M, Bonefeld-Dahl C, Bauer W, et al. The Assessment List for Trustworthy Artificial Intelligence (ALTAI). European Commission; 2020. p. 34. Available from: https://books.google.es/books?id=cu8dEAAAQBAJ&dq=The+Assessment+List+for+Trustworthy+Artificial+Intelligence+(ALTAI).&lr=&hl=es&source=gbs_navlinks_s

[52] Dignum V, Penagos M, Pigmans K, Vosloo S. Policy Guidance on AI for Children. Communications of UNICEF; 2021. Available from: https://www.unicef.org/globalinsight/reports/policy-guidance-ai-children

[53] Charisi V, Gomez E, Mier G, Merino L, Gomez R. Child-robot collaborative problem-solving and the importance of child's voluntary interaction: A developmental perspective. Frontiers in Robotics and AI. 2020;7:15

[54] Charisi V, Davison D, Wijnen F, Van Der Meij J, Reidsma D, Prescott T, et al. Towards a child-robot symbiotic co-development: A theoretical approach. In: AISB Convention 2015. The Society for the Study of Artificial Intelligence and the Simulation of Behaviour. The Society for the Study of Artificial Intelligence and the Simulation of Behaviour (AISB); 2015. Available from: https://dspace.library.uu.nl/handle/1874/324608

[55] Charisi V, Dennis L, Fisher M, Lieck R, Matthias A, Slavkovik M, et al. Towards moral autonomous systems. arXiv. 2017. Available from: https://arxiv.org/abs/1703.04741

[56] Charisi V, Merino L, Escobar M, Caballero F, Gomez R, Gómez, E. The effects of robot cognitive reliability and social positioning on child-robot team dynamics. In: 2021 IEEE international conference on robotics and automation (ICRA). IEEE; 2021. pp. 9439-9445. Available from: https://ieeexplore.ieee.org/abstract/document/9560760?casa_token=hFdtPyqOh58AAAAA:ZLONvEfeRPancjL0fxyjWGoMl97LqW8kafeUE8QDKgoaDPyzJz4XYGuDfhcvy-sJMWTgZJfq8sP2G-s

[57] Escobar-Planas M, Charisi V, Gomez E. "That robot played with us!" children's perceptions of a robot after a child-robot group interaction. Proceedings of the ACM on Human-

Computer Interaction. 2022;**6**(CSCW2): 1-23

[58] Linstone HA, Turoff M, et al. The Delphi Method. Reading, MA: Addison-Wesley; 1975

[59] Kovačević N. Aleksandra Stojiljković, and Mitar Kovač application of the matrix approach in risk assessment. Operational Research in Engineering Sciences: Theory and Applications. 2019;**2**(3):55-64

[60] Marina Escobar-Planas, Emilia Gómez, Carlos-D Martınez-Hinarejos. Enhancing the Design of a Conversational Agent for an Ethical Interaction with Children. Proc. IberSPEECH 2022. 2022:171–175. DOI: 10.21437/IberSPEECH.2022-35